eXactagon

by

David Biagini

eXactagon is the culmination of my occasional explorations into a single unifying method of utilizing only a compass and a straightedge, along with a drawing and paper medium, to create any regular polygon.

I truly hope that you also enjoy this aha moment.

Contents

 Foreword

I remember being told in Geometry class that there was no singular way to accurately draw any regular polygon with only a straight edge and a compass. Ever since I have always felt there must be, and in 2010 I worked out a good general method I dubbed Perfectagon.

At the same time I obtained perfectagon.com to showcase the method that I had included in a product development proposal called Puzzle Postage that was submitted to the United States Postal Service in that same year.

Shortly after perfectagon.com went live there was one very kind supporter who sent a message regarding an alternate solution for a nine sided regular polygon also known as an enneagon, but still no all encompassing solution.

Since then to the date of this publication, I removed myself from the social network. Subsequently coming to the decision that I would no longer seek any other outside solution, so that if one exists I would eventually enjoy the success of making the discovery on my own.

Having finished an extensive round of marketing for Store Wears Episode Fore and Hello Hummingbirds, instead of resuming the current development of several screen plays, I decided to revisit an unfinished project. Such is what has resulted in this amazing book.

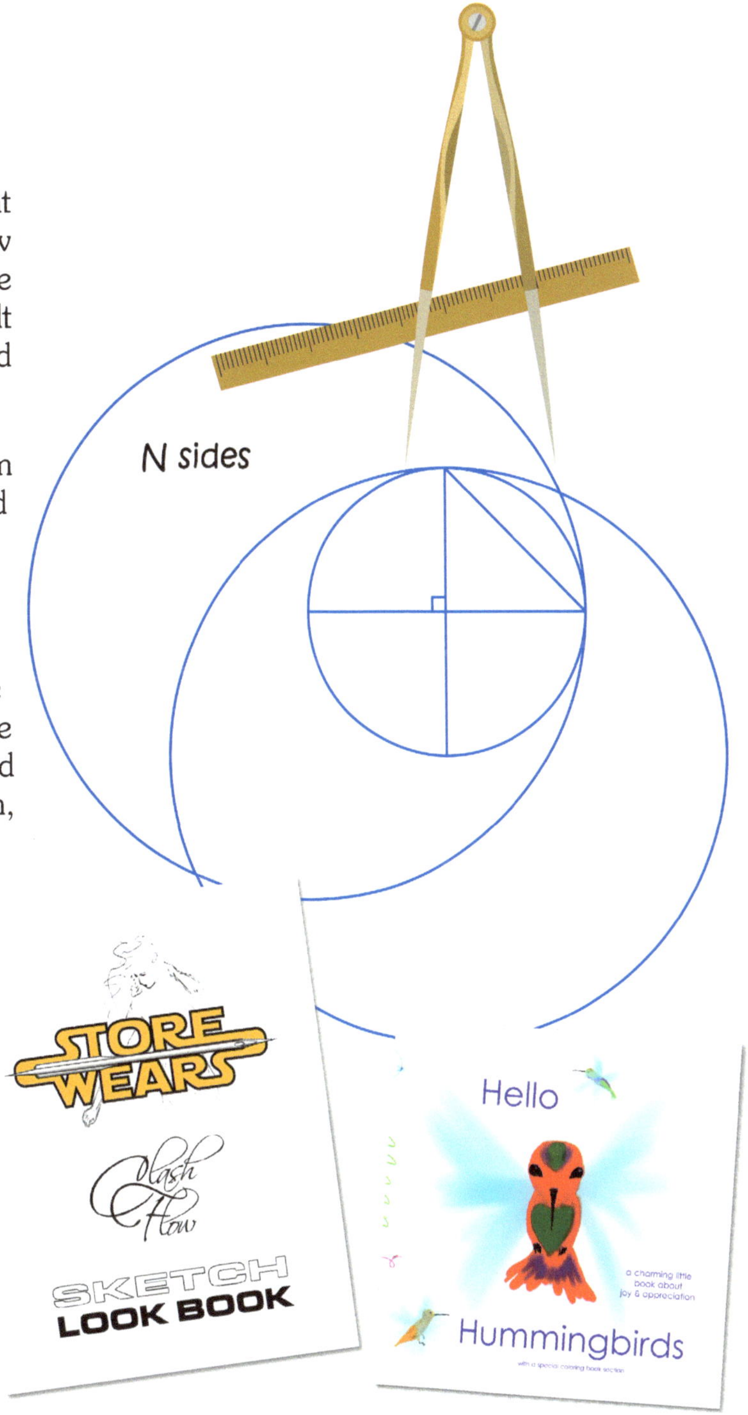

N sides

Introduction

Through countless hand drawn renditions I have realized that using even the most exacting drafting elements may still produce inaccuracy. Any misplacement of measures, no matter how minute, will likely compound into radical shifts and gaps where two points are expected to coincide.

With a relaxation of expectation my ultimate goal was to discover a relatively easier method than Perfectagon that would render more accurate results. So once again I took up the task of solving the eXactagon Method.

The usual course was; have an idea, scratch out some rough comps, utilize CAD to test, find out there needs to be a new idea.

For ease of set up I decided to simply use the diameter of a single small base circle as the determining factor for every other circle so that a circle with a diameter of three bases would be exactly half of a circle of six bases. With the aid of CAD I included the regular polygon within each associated circle, so that a three base circle would have a trigon, a four base would have a tetragon, a five base a pentagon, and so on.

To further the efficiency I chose to align the series of increasing diameter circles along one side at what would essentially be a single point. All polygons would have one side parallel closest to the point of alignment which happens to form a beautiful blossoming effect.

Back with paper and a compass and straightedge, I discovered that a six base circle drawn from a key point of the hexagon intersected with enough precision to not only easily create the usual hexagon, but also a pentagon along the five base circle and a heptagon along the seven base circle. Beyond heptagon the method started to falter.

Yet a bit more playtime revealed that a straight line from a specific quadrant of the six base circle through the same hexagon key point rendered a nice precision mark for creating enneagon through tridecagon on the respective base circles.

Further exploration relayed that a straight line from a specific quadrant of the twelve base circle through a similar key point of the dodecagon developed an even more precise construction for polygons of more sides.

Although fortified with this new more efficient Align Method of eXactagon, which shifted the project of discovery to the project of formulating a system to illustrate the process, I still kept trying some other variations of circle and polygonal relationships.

Behold the Ratio Method of eXactagon. It is by far the most exacting general method I have been able to formulate.

The following illustrations will demonstrate the Align Method and the Ratio Method of eXactagon. Additionally included are reference illustrations to the Perfectagon Method as well as original imagery from perfectagon.com, and a few choice surprise renderings.

Utilizing Standard Methods it is incredibly straight forward to create a trigon, tetragon, hexagon, and octagon. In fact the magic of the hexagon creation is a key factor in how the

eXactagon methods are brilliant for the creation of any other regular polygon.

Individual demonstration graphics are presented with a tiered color structure. A reference color key is included for ease of interpreting step order.

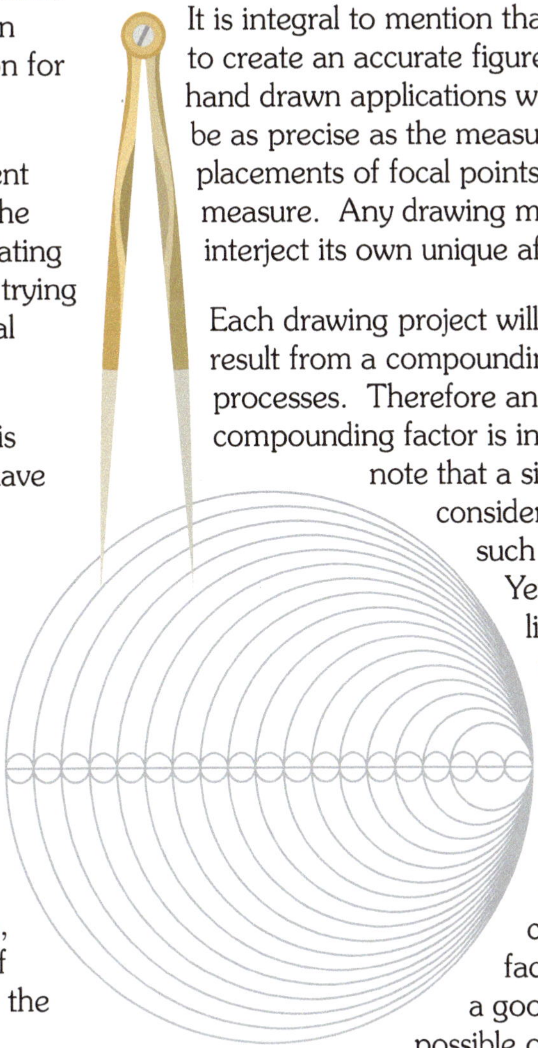

It is integral to mention that attempts to create an accurate figure utilizing hand drawn applications will only be as precise as the measures and placements of focal points for any measure. Any drawing media will also interject its own unique affects.

Each drawing project will automatically result from a compounding of consecutive processes. Therefore an approximate compounding factor is indicated. Please note that a single action is considered one factor, such as drawing a line. Yet even a straight line could be more than one factor.

For image creation that spans multiple illustration sets, summing the compounding factors should give a good indication of possible outcome.

Step Ordering

Later

Earlier

Scope Of Measure

Although the concept of using only a compass and straightedge to draw any regular polygon may seem obvious, in actuality the opposite may be more true.

With the complete inclusion of electronics at the start of a human life, for games, learning, mathematical computation, graphing and drawing as well as the means to have practically all expression of basic creativity stripped away by any hand held communication device, there is a diminishing of incentive to explore tangibly.

A requirement for measuring is the ability to contrive the distance between two points. So a compass could be as simple as a pin with a string. For the purpose of eXactagon a compass is a device with two straight legs pivoted at one end such that the legs may be securely positioned at any angle enabling the ability to obtain and reproduce a measurement. One leg should have a fine tip to be utilized as the first point of measure while the other leg has a fine tip drawing medium for ease of marking the

second point of any measure as well as the creation of a circle with a radius of the set measure.

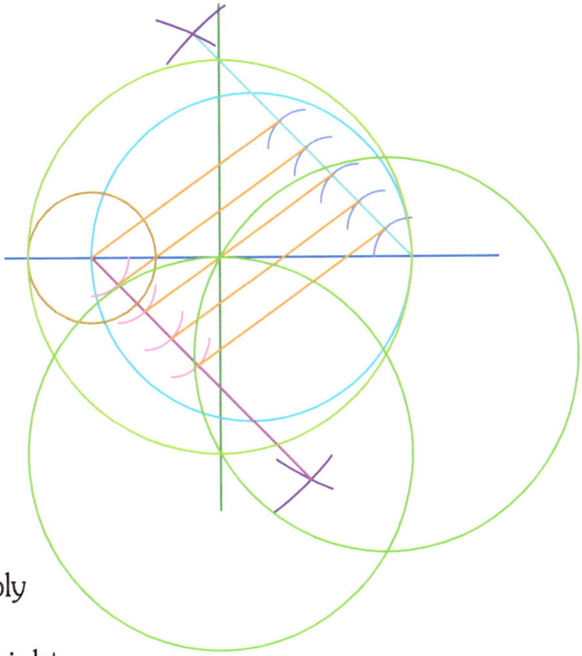

The
straightedge
just needs to
be considerably
close to
absolutely straight,
and does not need any
form of unit measure.

A normal method for marking any
measure with a compass is to place
the fine tip at the first point and the
drawing tip at the second point.

This type of measuring can effectively
form a circle from any chosen focal
point or origin. Most of the illustrations
presented indicate the entire measured
circle rather than the arc portion of intersect.

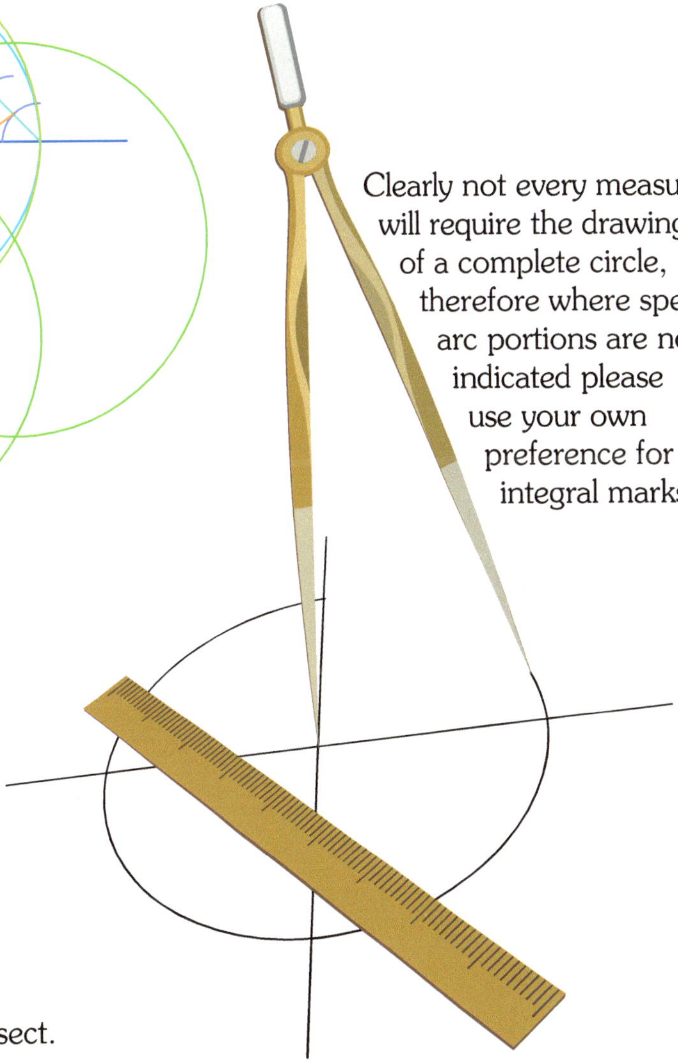

Clearly not every measure
will require the drawing
of a complete circle,
therefore where specific
arc portions are not
indicated please
use your own
preference for
integral marks.

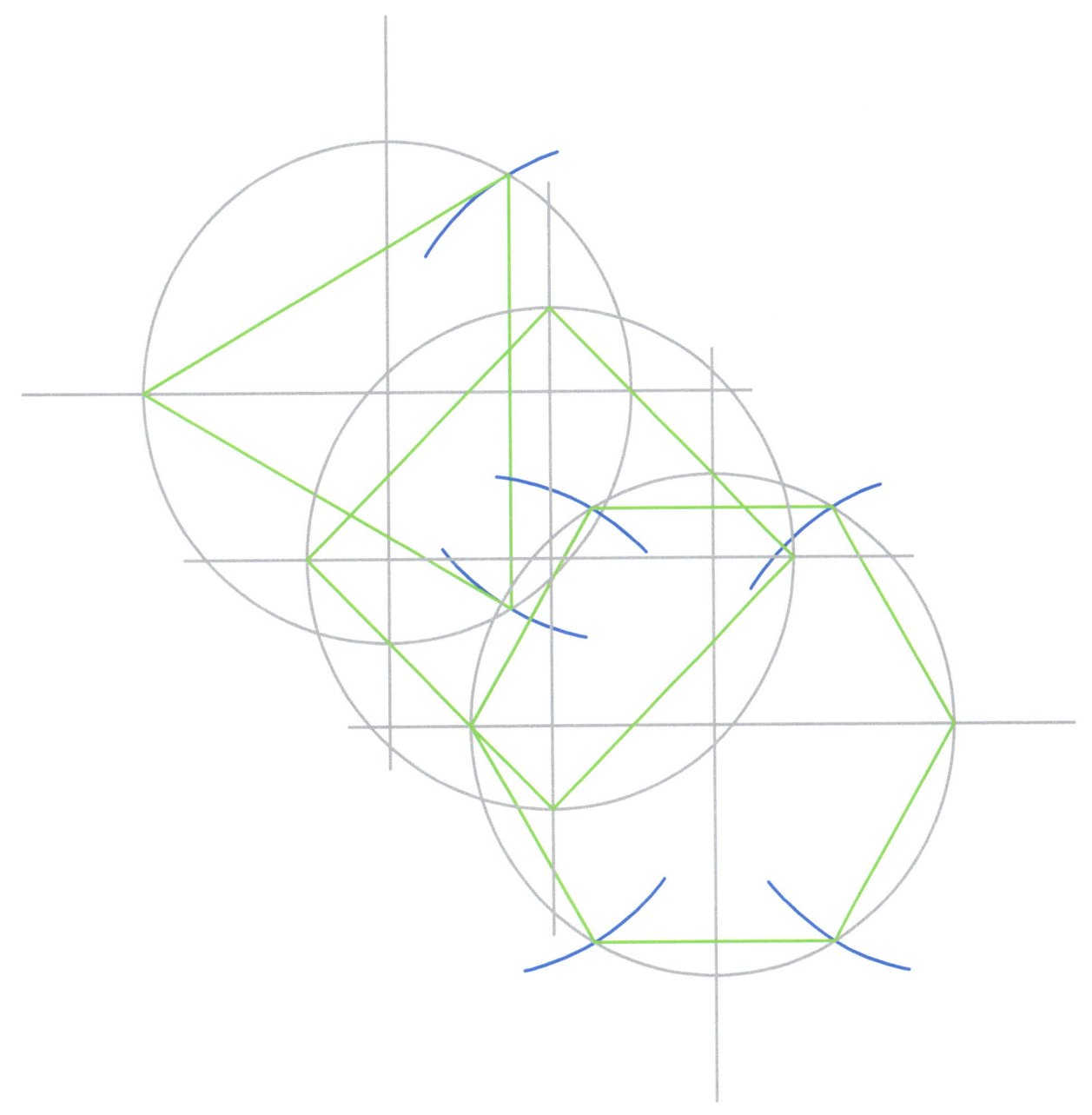

Standard Methods

Quadrant Layout

Any references made to portions of a circle will relate to the quadrant layout which utilizes the traditional use of the Roman Numerals, I, II, III & IV.

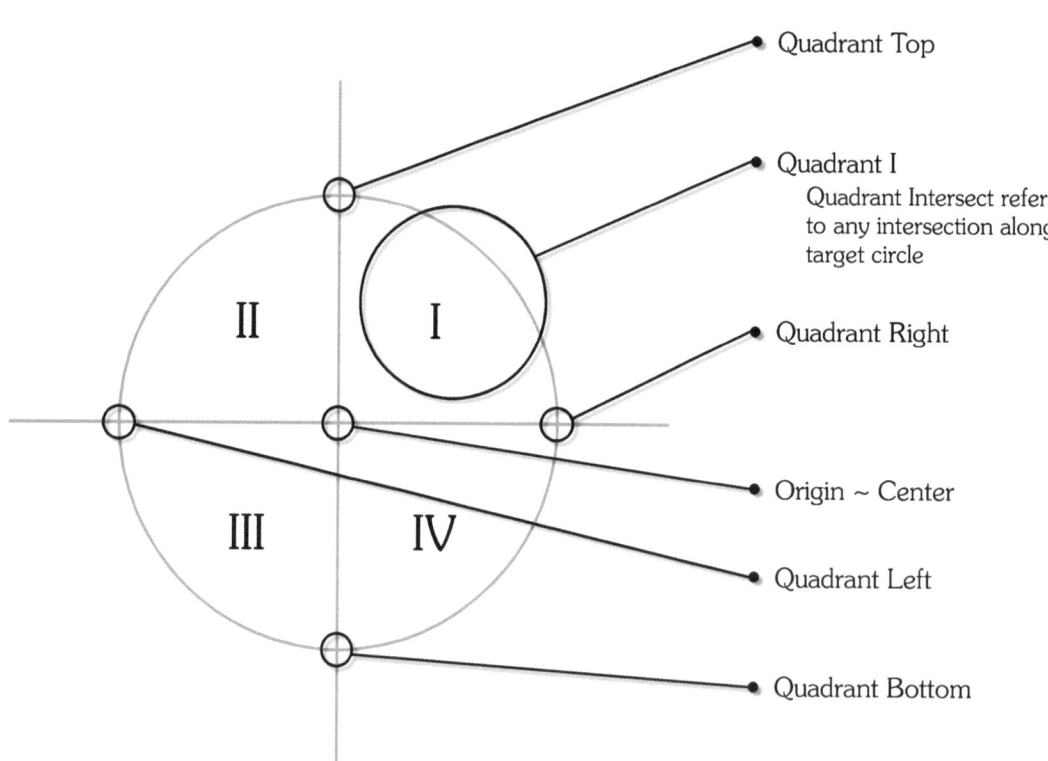

Quadrant Top

Quadrant I
  Quadrant Intersect refers to any intersection along target circle

Quadrant Right

Origin ~ Center

Quadrant Left

Quadrant Bottom

Step Ordering

Later

Earlier

Circle With Center Axes

a.

c.

b.

d.

e.

a. draw a straight line

b. two circles spaced apart along line

c. utilize intersecting circles to bisect line, resulting in two perpendicular axes

d. set origin of target circle at axes intersection

e. a good practice is to create a separate clean version utilizing all obtained measurements, that may only require a circle and one or two indications of a quadrant intersect

Standard Methods

Divide Line Equally

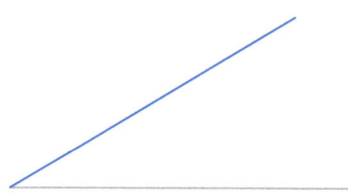

a.

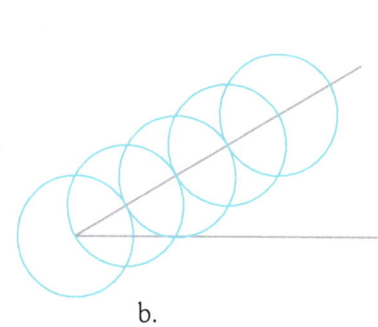

b.

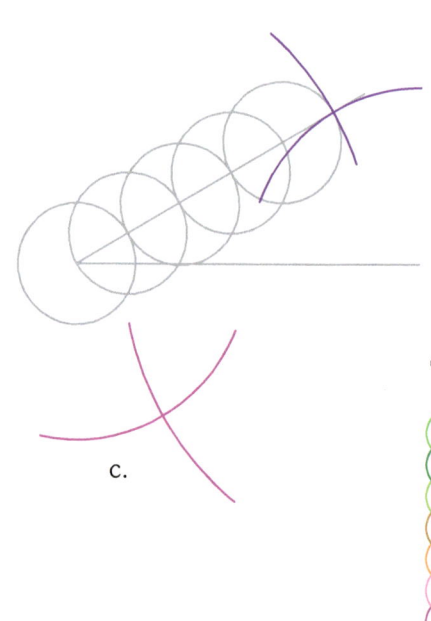

c.

Step
Ordering

Later

Earlier

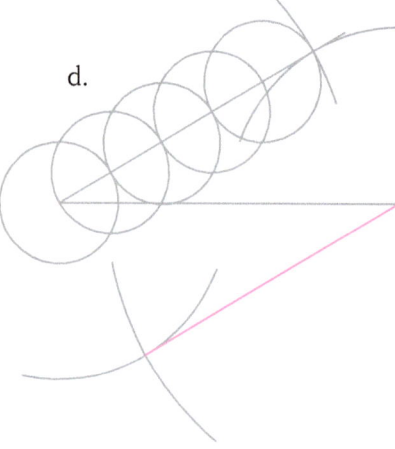

d.

e.

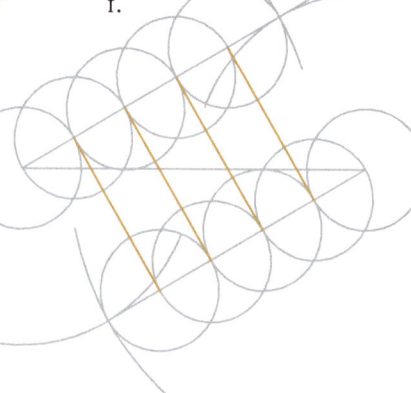

f.

a. angled reference line over target line

b. divide reference line into equal lengths

c. determine endpoint of reference line and mark opposite target line

d. draw second reference line

e. mark identically to first reference line

f. connect reference marks

g. resulting in target line divided equally

g.

3 Sided ~ Trigon

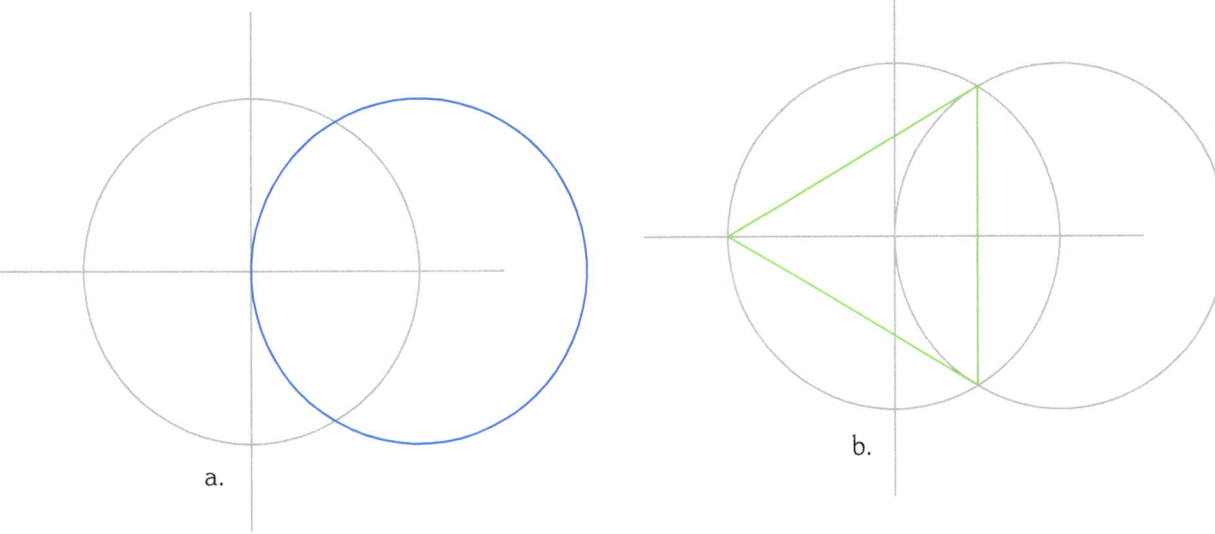

a.

b.

a. duplicate target circle centered at quadrant right

b. connect quadrant left with each intersection of the two circles

c. resulting in three sides of a trigon

All polygonal names referenced are in common use and derived predominantly from Greek.

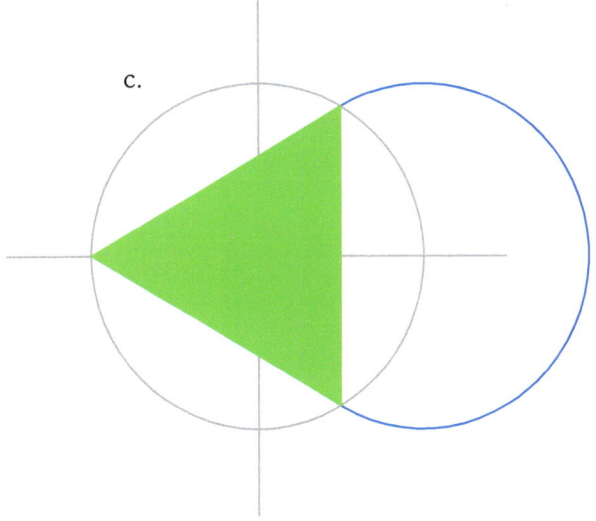

c.

Standard Methods

4 Sided ~ Tetragon

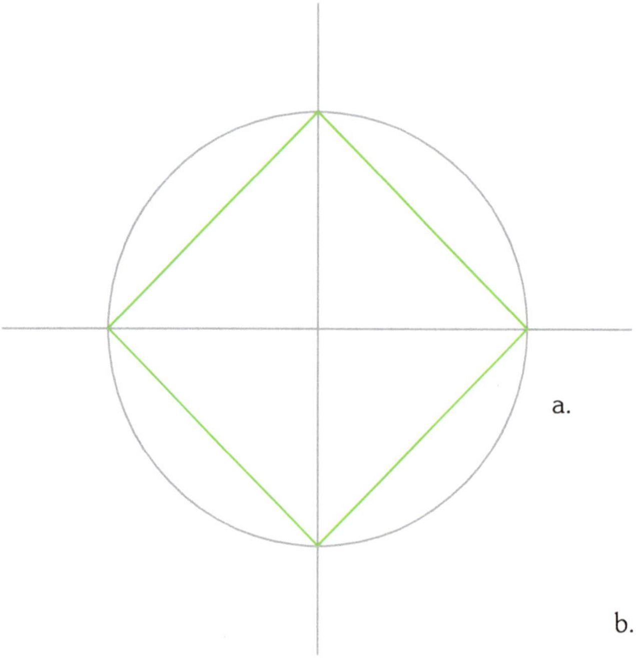

a.

b.

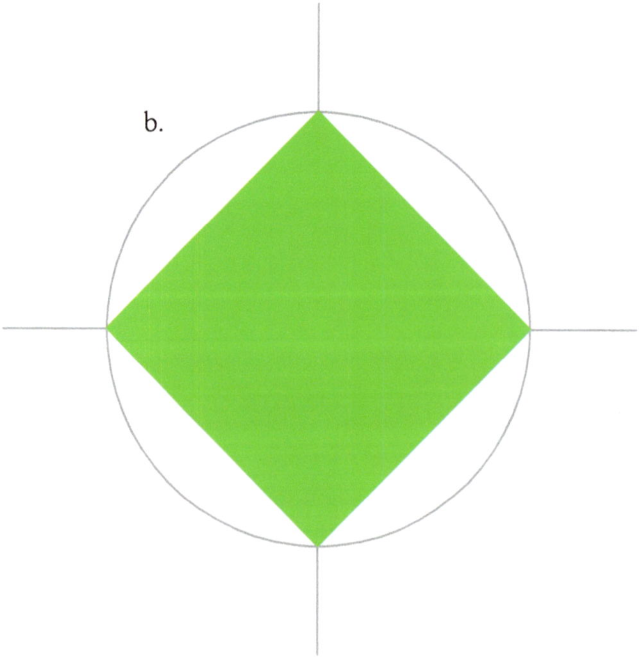

Step
Ordering

Later

Earlier

a. connect quadrant right, top,
left and bottom

b. resulting in a tetragon

6 Sided ~ Hexagon

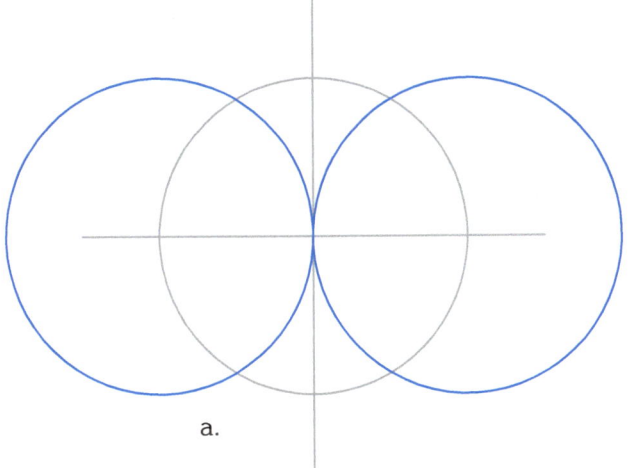

a.

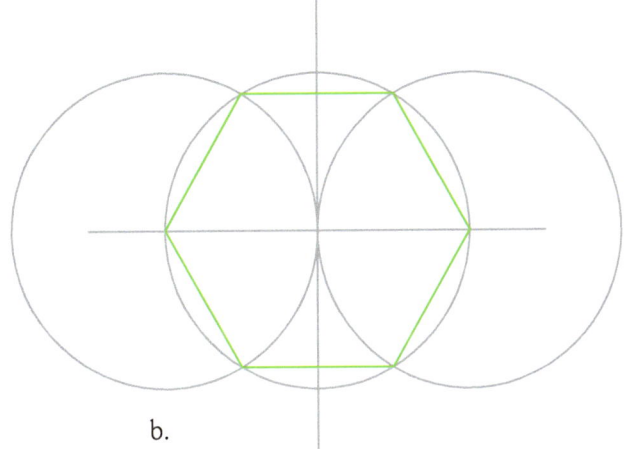

b.

a. extend step -a- of the trigon with a duplicate circle at quadrant left

b. connect quadrant right, quadrant I intersect, quadrant II intersect, quadrant left, quadrant III intersect, and quadrant IV intersect

c. resulting in a hexagon

c.

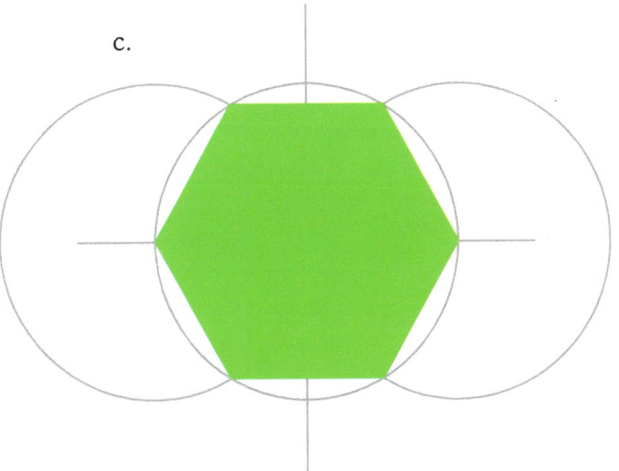

Standard Methods

8 Sided ~ Octagon

Bisecting a given side of any regular polygon will effectively enable the creation of a new regular polygon with twice the number of sides.

a. center a circle at quadrant right and an equal circle at quadrant top to bisect quadrant I

b. from quadrant right measure to quadrant I bisection, then intersect quadrant IV

c. create identical intersections for quadrant II and III

d. connect the eight key points

e. resulting in an octagon

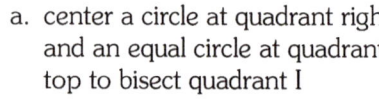

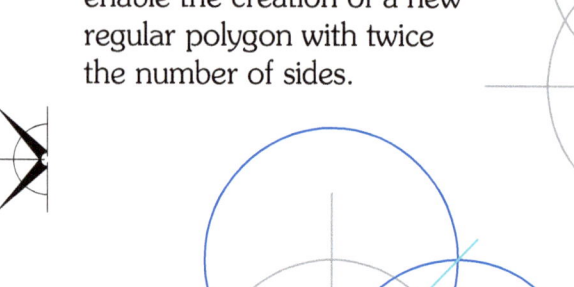

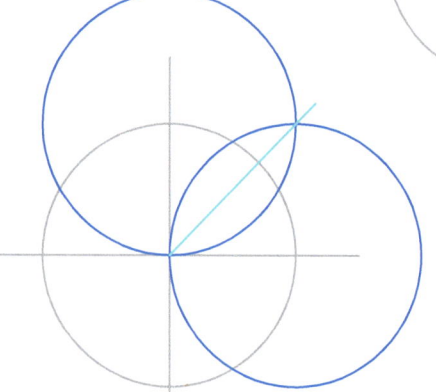

a.

b.

Step
Ordering

Later

Earlier

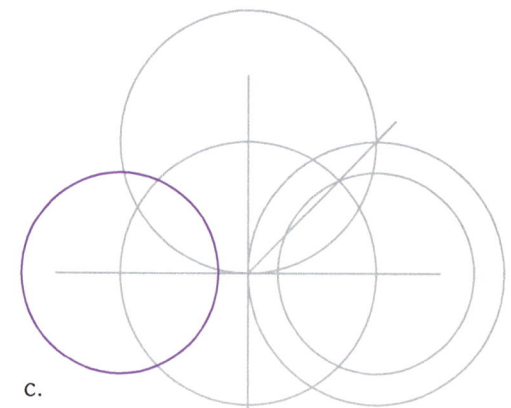

c.

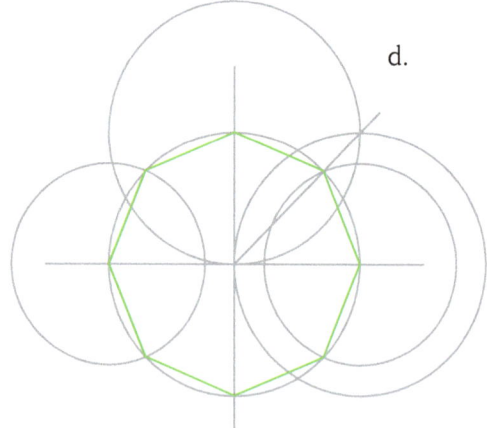

d.

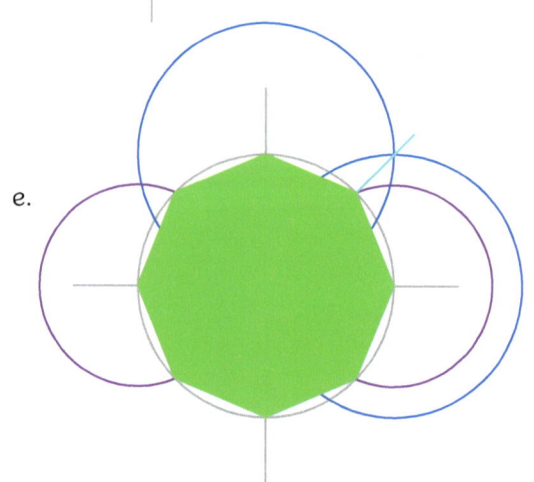

e.

16 Sided ~ Hexadecagon

a. extend step -a- of the octagon to bisect half of quadrant I

b. from quadrant right measure to resulting quadrant I intersect, then intersect quadrant IV

c. create identical intersects for all quadrants by centering measured circle at quadrant top, left and bottom

d. connect the sixteen key points

e. resulting in a hexadecagon

a.

b.

c.

d.

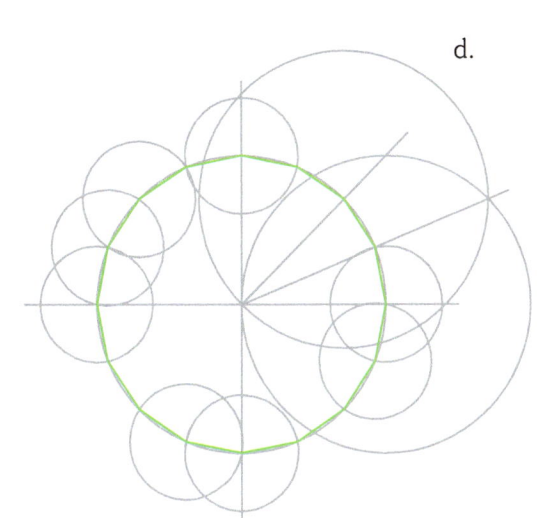

e.

eXactagon Align Method

eXactagon Align Method

Primer

The radius for the base circle is a unit of one, and all other shown measurements and ratios are in reference to that base unit.

1.000

When reviewing the figures throughout this presentation, please be aware that slight anomalies are apparent in dimensioning and snap created precision. Such as theoretically identical side measurements not matching.

The CAD software still produced remarkable results, and any variations should easily be within the realm of human hand drawn capabilities.

Both the eXactagon Align Method and Ratio Method rely on incrementally increasing circle diameters relating to a single base measure indicated as a base circle with a radius matching a unit of one.

The Align Method is shown here to the right, yet it should be identically effective with any orientation, as long as all related measures are adjusted to relate as if transformed from the demonstrated figures.

An integral element of the Align Method is that each created polygon is situated with one side parallel to the one side of every other polygon that is closest to the aligned point of the incremented circles.

As demonstrated in the Standard Methods, the one polygon that elegantly blossoms from any circle is the hexagon. The gifts of the magic six sided regular polygon are further shared with every other when used within an incremental system. Similarly the dodecagon extends the gifts of the hexagon, which appears to be consistently paid forward with the doubling of each stepped hexagonal figure.

19 18 17 16 15 14 13 12 11 10 9 8 7 6 5 4 3

eXactagon Align Method

Base 6 Arc

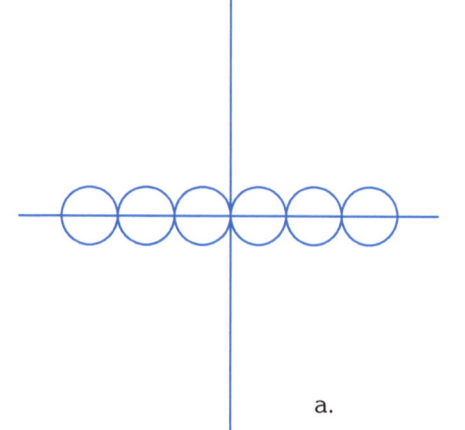

a.

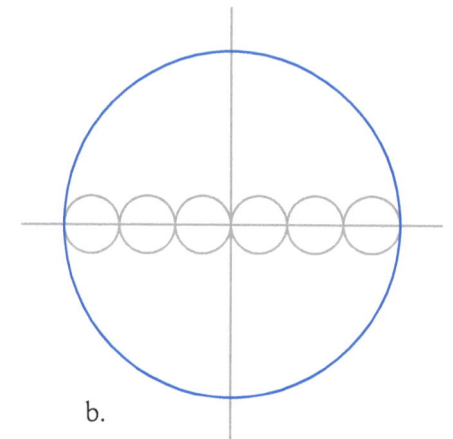

b.

c.

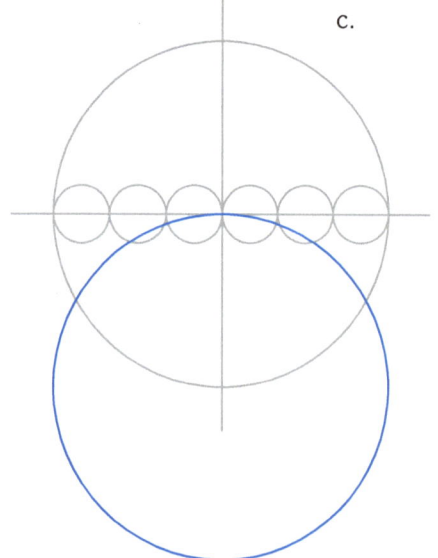

d.

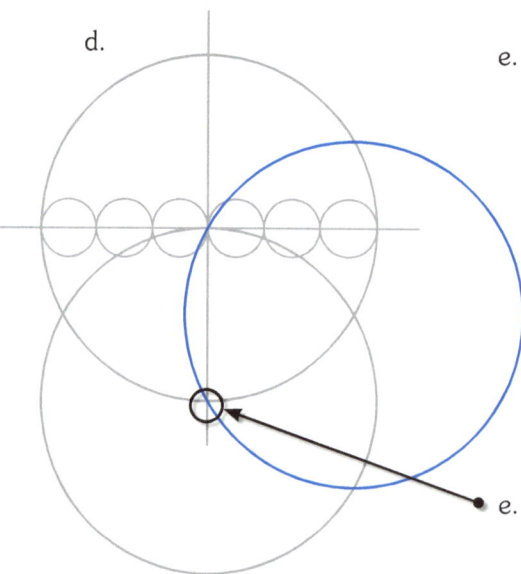

a. using a base circle with a radius of one unit, measure out six base circle diameters along a straight line, include a perpendicular center axis

b. create a circle matching a diameter of six base circles

c. place a duplicate circle centered at quadrant bottom

d. place a duplicate circle centered at the newly created quadrant IV intersect

e. the arc intersecting quadrant bottom is the base 6 arc

Step Ordering

Later

Earlier

e.

Base 6 Line

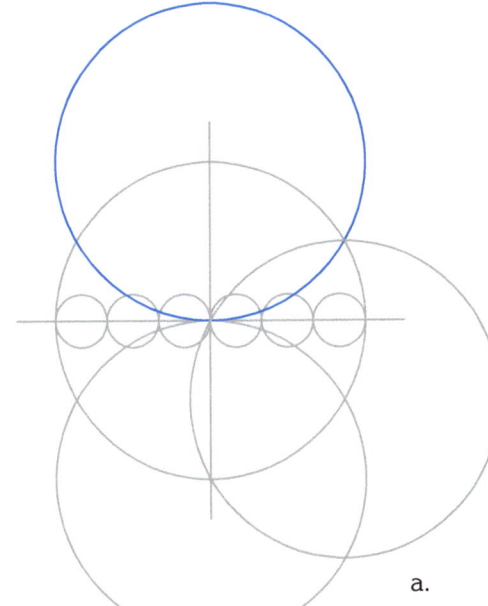

a.

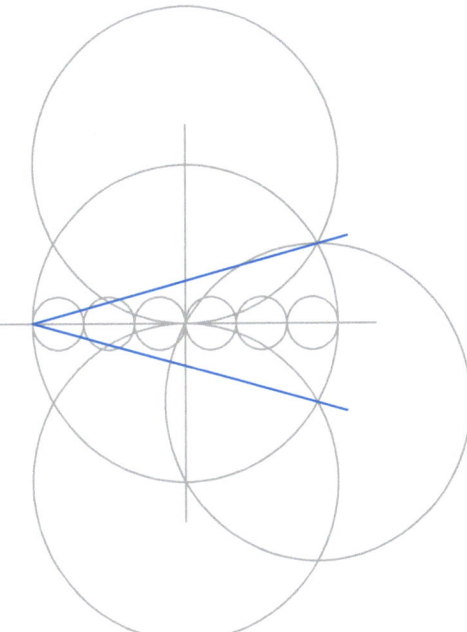

b.

c.

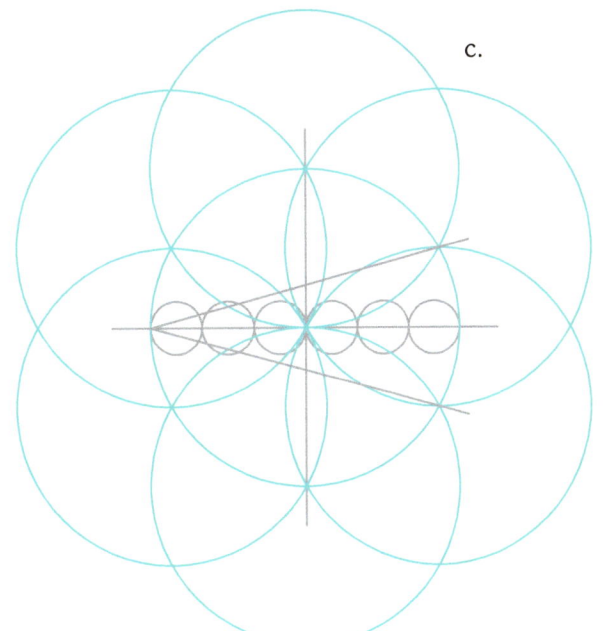

a. to help ensure accuracy
   of quadrant I intersect,
   extend the base 6 arc
   setup with an additional
   duplicate circle centered
   at quadrant top

b. draw two lines; a line
   from quadrant left
   through the quadrant I
   intersect, and a line
   from quadrant left
   through the quadrant IV
   intersect; the intersection
   of these two lines
   through each aligned
   circle will determine the
   side measure for the
   associated polygon

c. known as the flower
   of life, the symbol is
   appropriately related
   with eXactagon

eXactagon Align Method

5 Sided B6 Arc ~ Pentagon

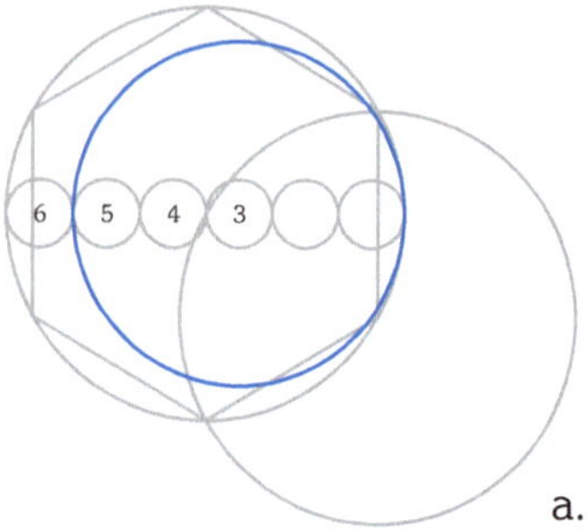

a.

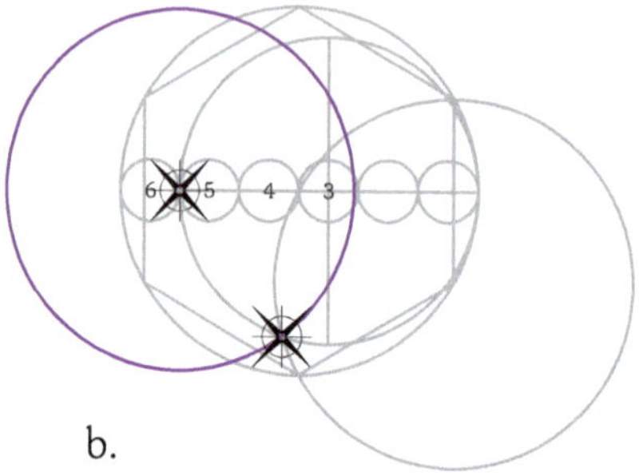

b.

c.

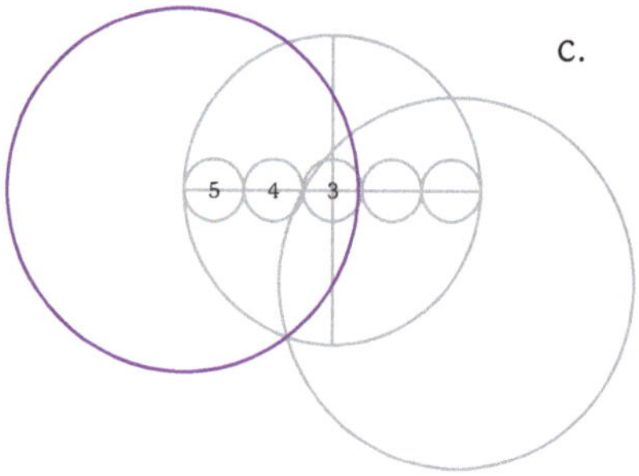

d.

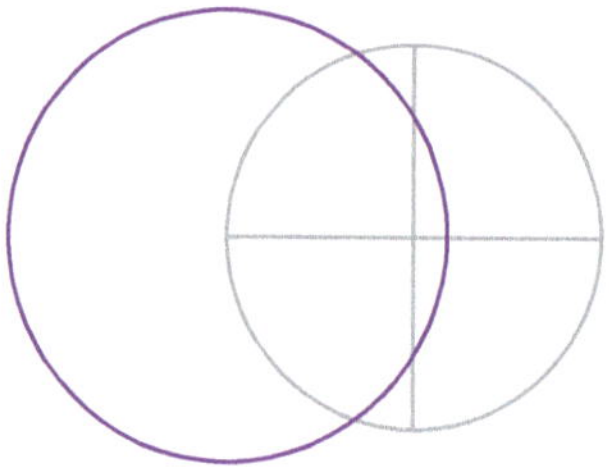

e.

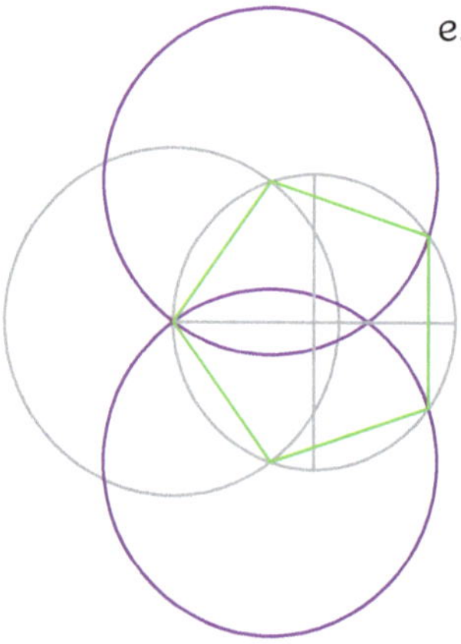

f.

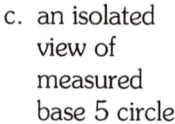

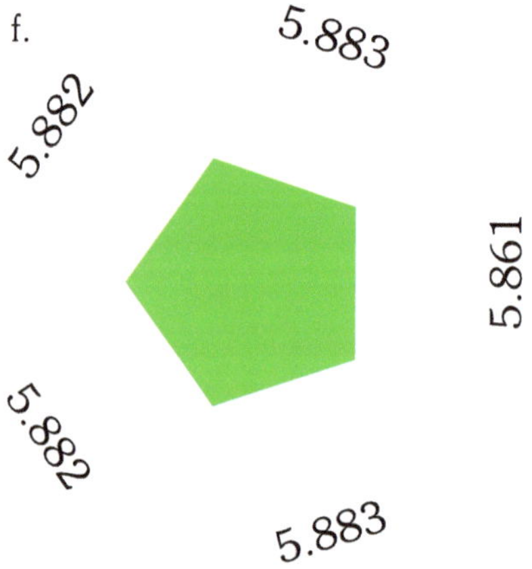

5.883

5.882

5.861

5.882

5.883

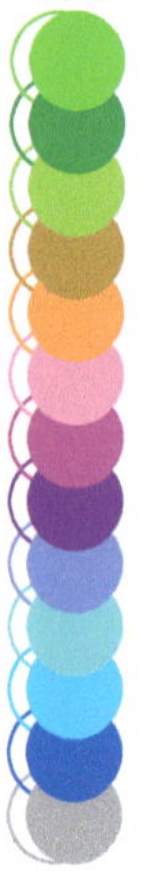

a. aligned to quadrant right of base 6 circle, make a base 5 circle, which is a circle of 5 base circle diameters

b. the base 6 arc indicates the pentagon side measure for the associated circle with a diameter of five base circles

Step Ordering

Later

c. an isolated view of measured base 5 circle

d. the resulting intersects in quadrant II and III indicate the centers to measure two more sides, with the last side self determined

Earlier

e. continuing the same measure will render the remaining sides

f. an indication of accuracy for pentagon

7 Sided B6 Arc ~ Heptagon

a. make a base 7 circle aligned to quadrant right

b. the base 6 arc indicates a two side measure for the heptagon that needs to be bisected extending a line from the center of the base 7 circle

c. measuring from quadrant left to quadrant III intersect is a single side

d. extending the measure through quadrant II is another.

e. continuing the same measure at each quadrant intersect will render the remaining five sides

f. the accuracy of the completed heptagon

a.

b.

c.

d.

e.

f.

6.080

6.079

6.080

6.047

6.080

6.080

6.079

6.080

eXactagon Align Method

9 Sided B6 Line ~ Enneagon

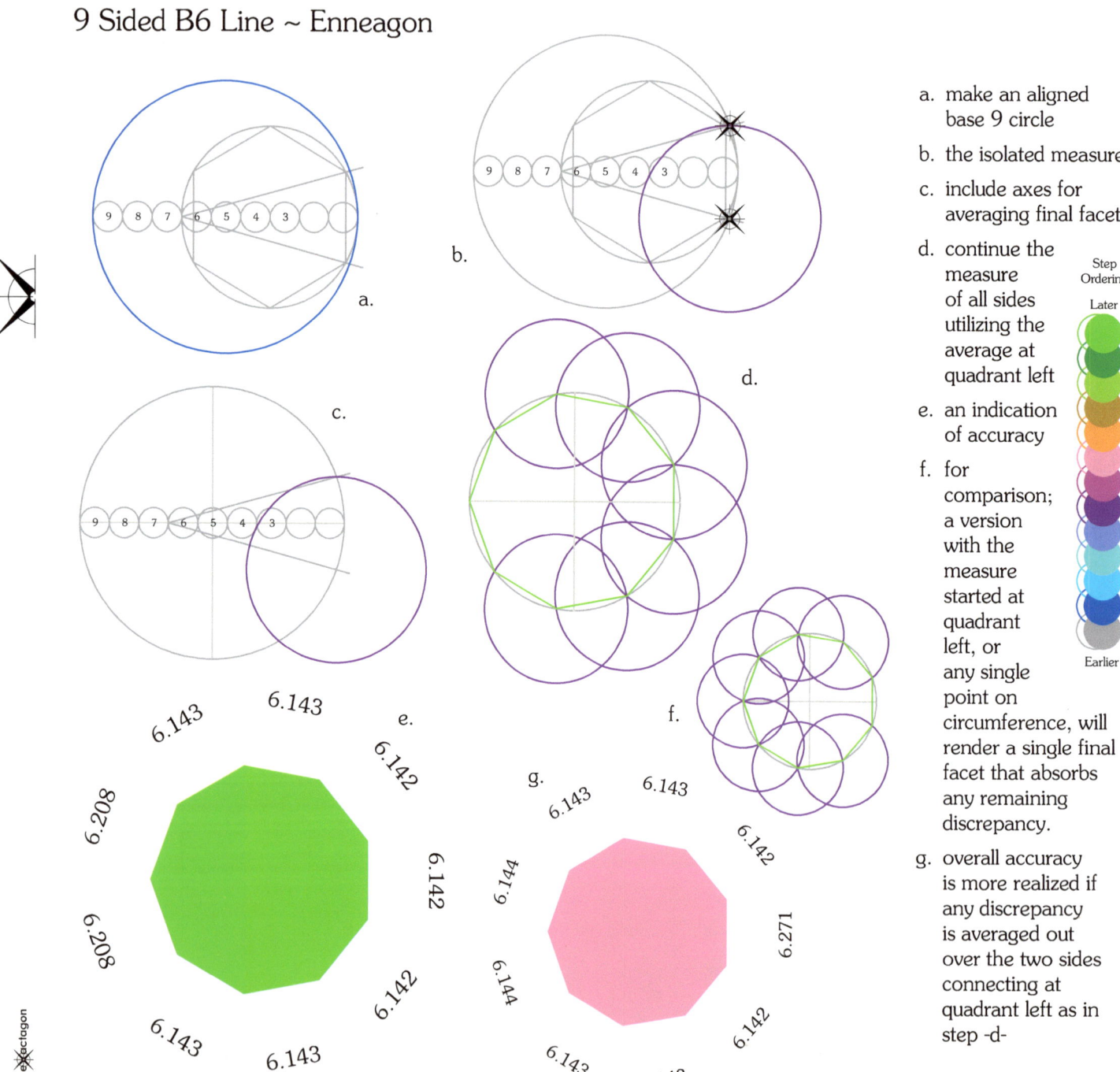

a. make an aligned base 9 circle

b. the isolated measure

c. include axes for averaging final facets

d. continue the measure of all sides utilizing the average at quadrant left

Step Ordering
Later

e. an indication of accuracy

f. for comparison; a version with the measure started at quadrant left, or any single point on circumference, will render a single final facet that absorbs any remaining discrepancy.

Earlier

g. overall accuracy is more realized if any discrepancy is averaged out over the two sides connecting at quadrant left as in step -d-

a.

b.

c.

d.

e.

f.

g.

6.143
6.143
6.143
6.142
6.208
6.142
6.208
6.142
6.208
6.142
6.143
6.143

6.143
6.143
6.144
6.144
6.143
6.143
6.142
6.271
6.142

28

11 Sided B6 Line ~ Hendecagon

a. make an aligned
   base 11 circle

b. the isolated
   measure

c. include axes for
   averaging final
   facets

d. continue the
   measure of all sides
   utilizing the average
   at quadrant left

e. an indication of
   accuracy

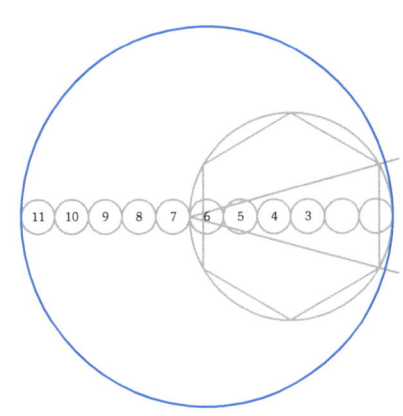

a.

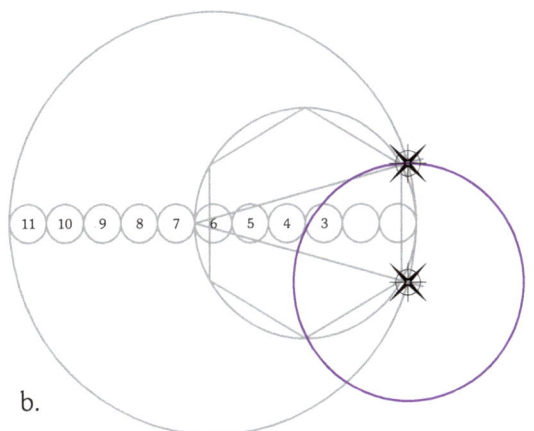

b.

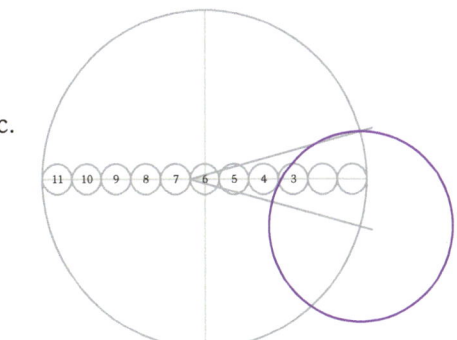

c.

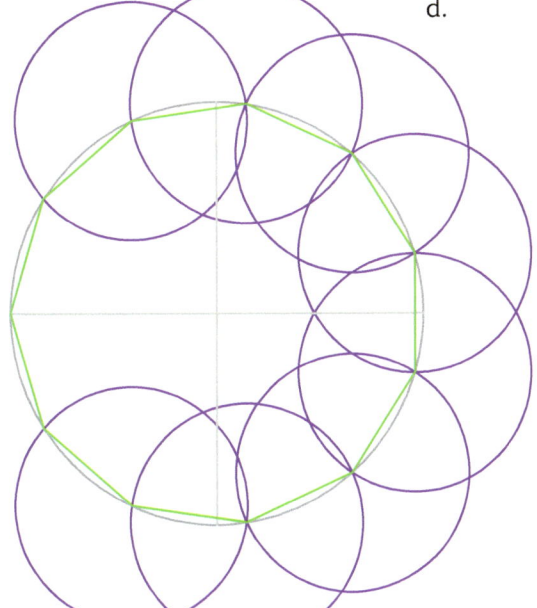

d.

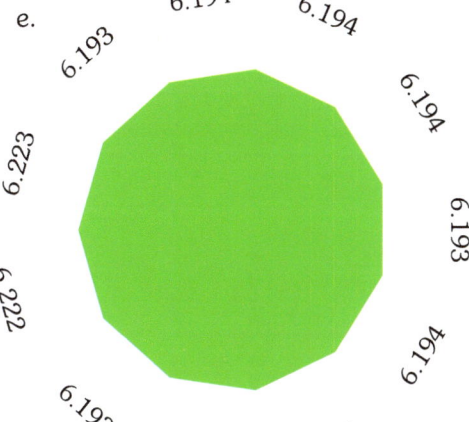

e.

6.194 6.194
6.193   6.194
6.223   6.193
6.222   6.194
6.193   6.194
   6.194 6.194

eXactagon Align Method

13 Sided B6 Line ~ Tridecagon

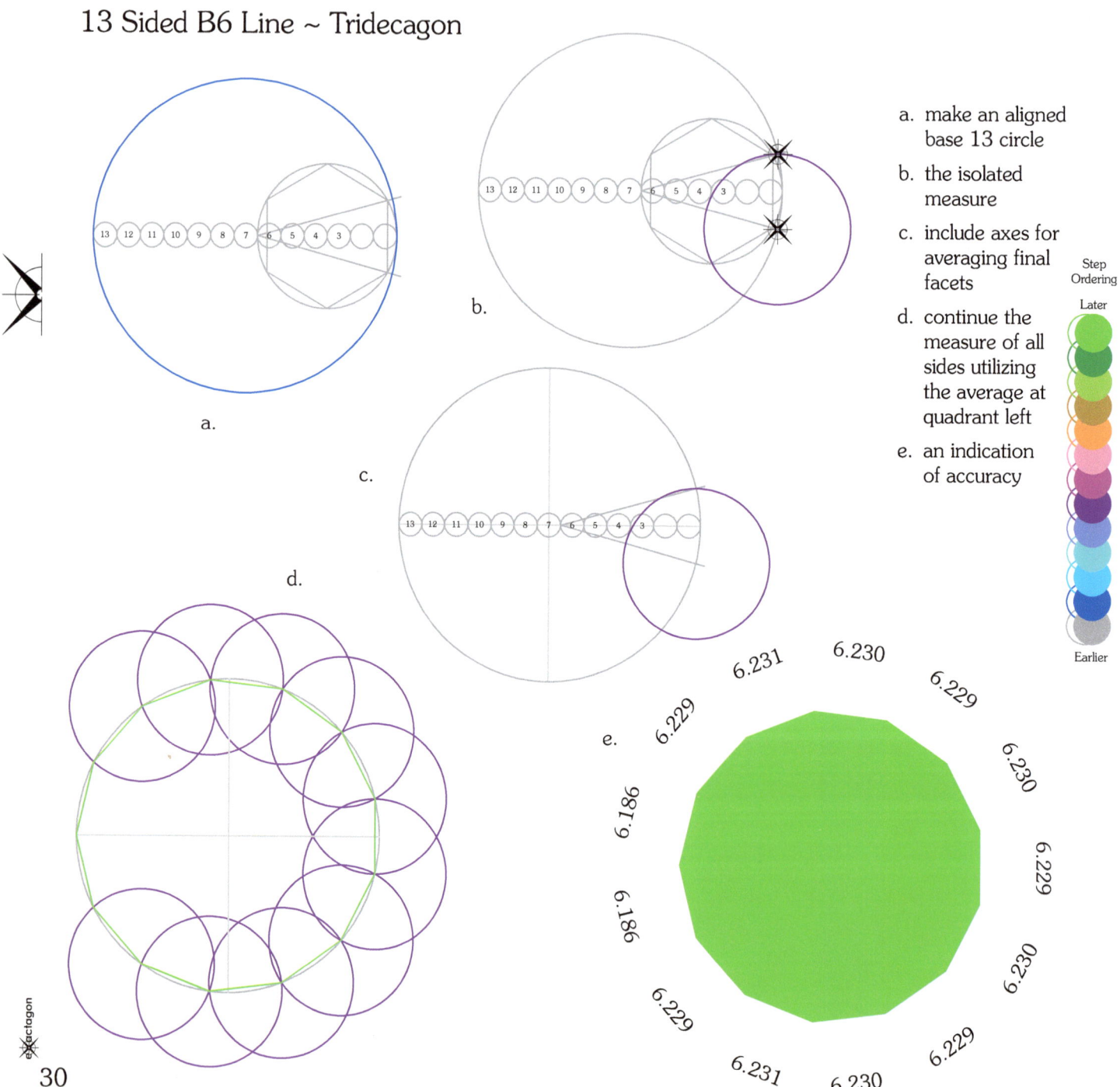

a.

b.

c.

d.

e.

a. make an aligned base 13 circle

b. the isolated measure

c. include axes for averaging final facets

d. continue the measure of all sides utilizing the average at quadrant left

e. an indication of accuracy

Step Ordering

Later

Earlier

6.230
6.231
6.229
6.229
6.230
6.229
6.230
6.186
6.186
6.229
6.229
6.231
6.230
6.230
6.229

eXactagon

15 Sided B6 Line ~ Pentadecagon

a. make an aligned base 15 circle

b. the isolated measure

c. include axes for averaging final facets

d. continue the measure of all sides utilizing the average at quadrant left

e. an indication of accuracy

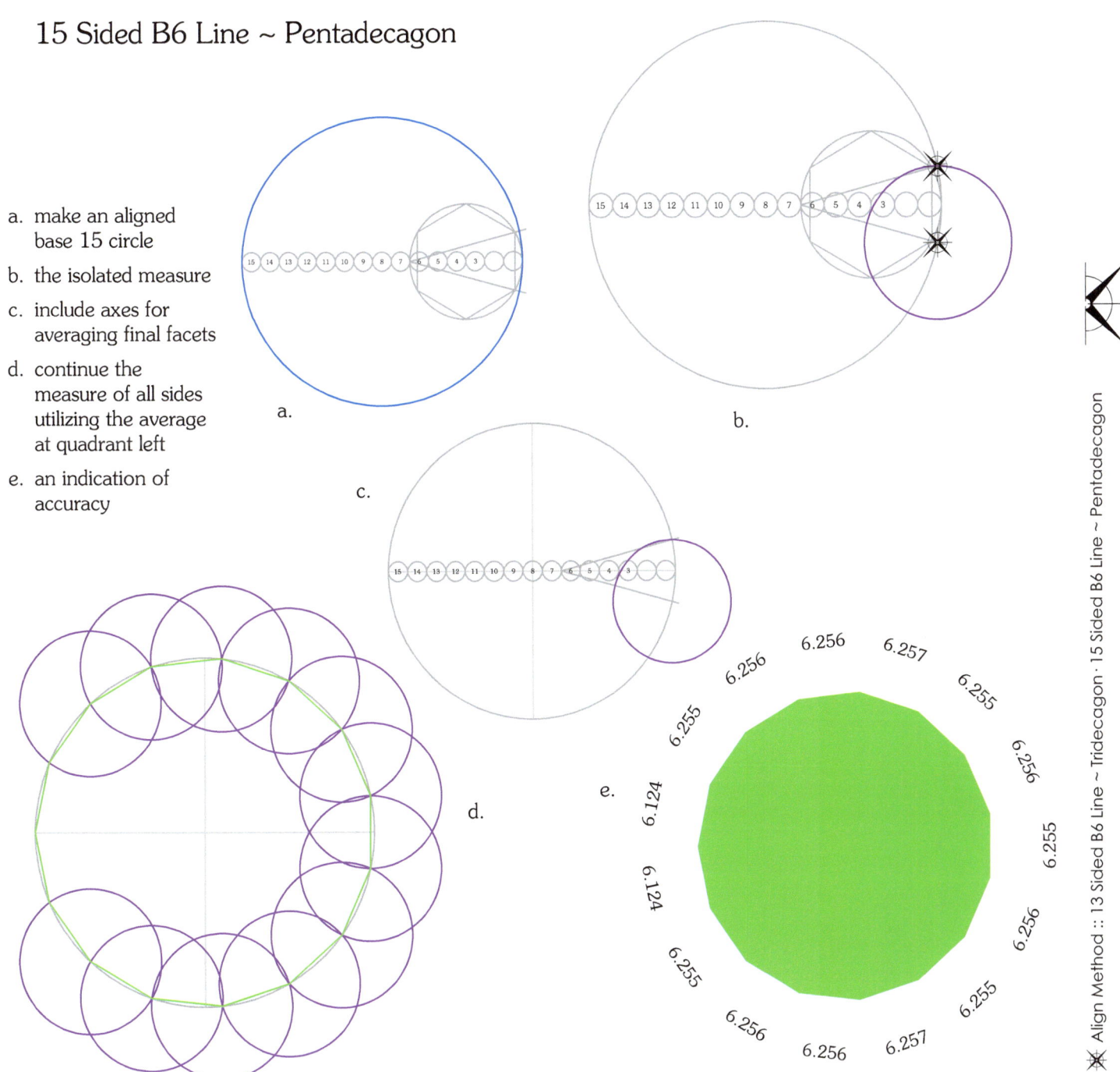

a.

b.

c.

d.

e.

6.256 6.256 6.257 6.255 6.256 6.255 6.255 6.255 6.256 6.257 6.256 6.256 6.255 6.255 6.124 6.124

eXactagon Align Method

17 Sided B6 Line ~ Heptadecagon

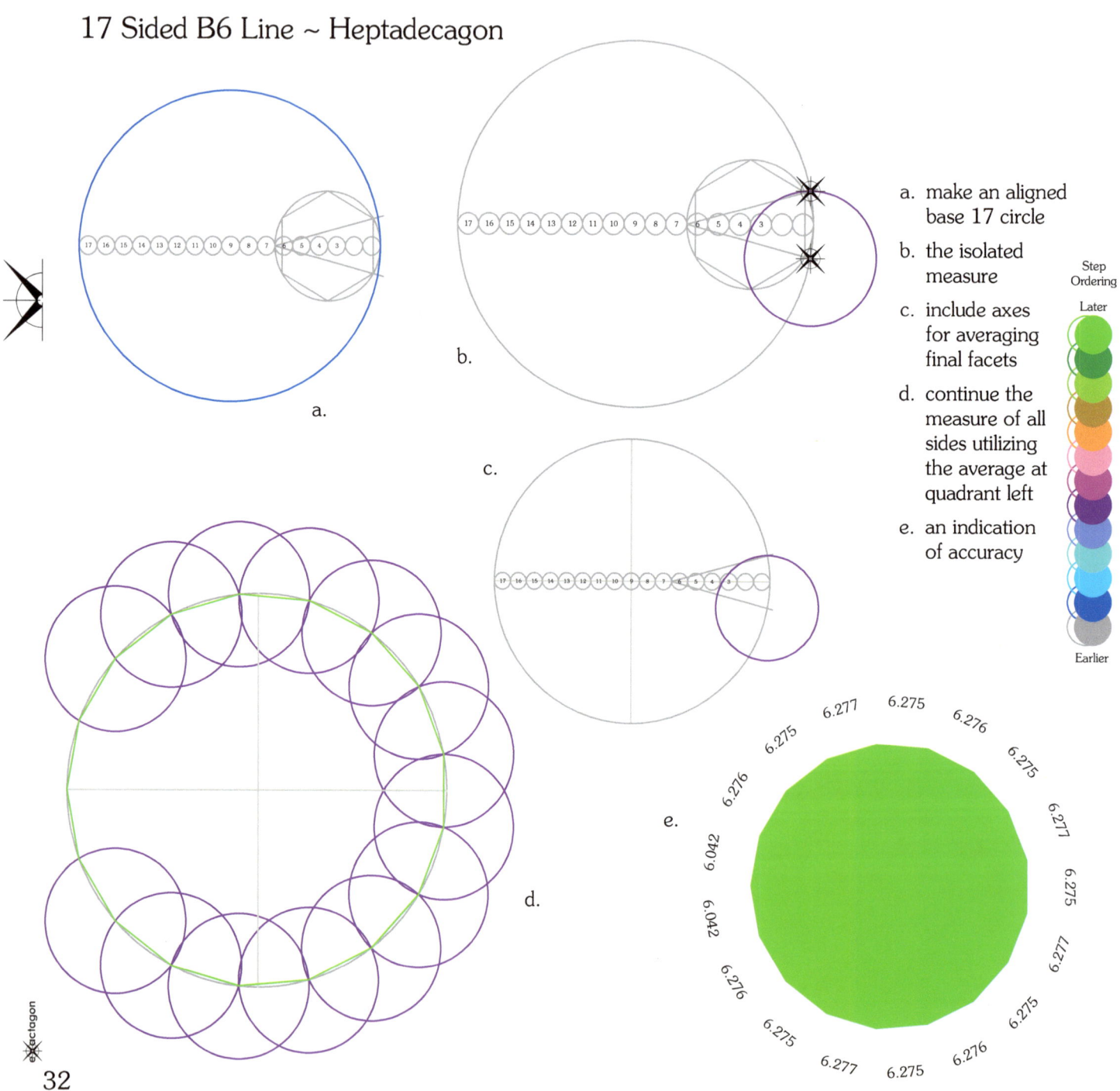

a.

b.

c.

d.

e.

a. make an aligned base 17 circle

b. the isolated measure

c. include axes for averaging final facets

d. continue the measure of all sides utilizing the average at quadrant left

e. an indication of accuracy

Step Ordering

Later

Earlier

19 Sided B6 Line ~ Enneadecagon

a. make an aligned base 19 circle

b. the isolated measure

c. include axes for averaging final facets

d. continue the measure of all sides utilizing the average at quadrant left

e. an indication of accuracy

a.

b.

c.

d.

e.

6.292 6.291 6.292 6.291 6.292
6.291 6.291
6.292 6.293
6.292 6.291
5.946 6.293
5.946 6.291
6.292 6.293
6.291 6.291
6.292 6.291 6.292 6.291 6.292

✳ Align Method :: 17 Sided B6 Line ~ Heptadecagon · 19 Sided B6 Line ~ Enneadecagon

Base 12 Line

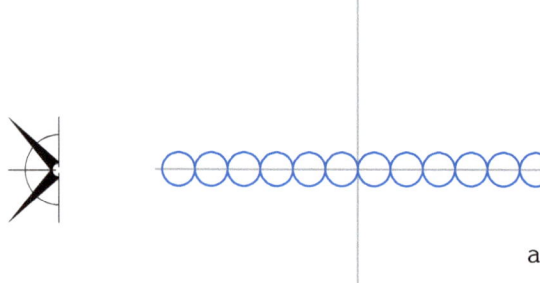

a.

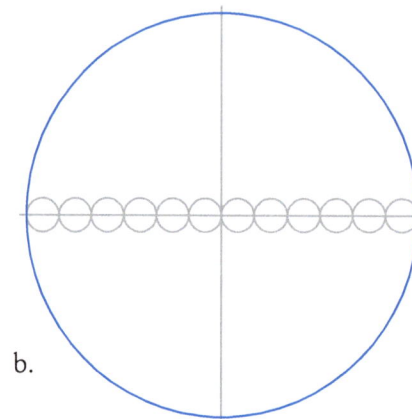

b.

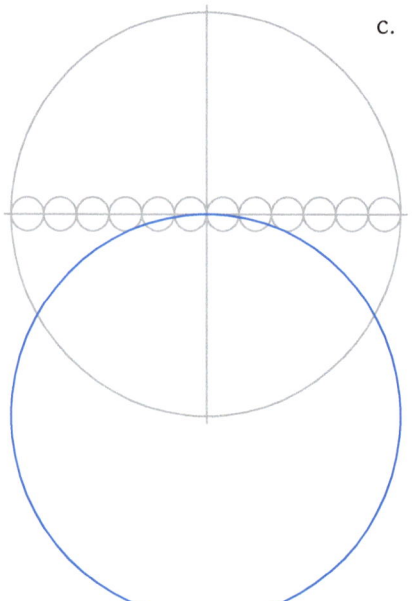

c.

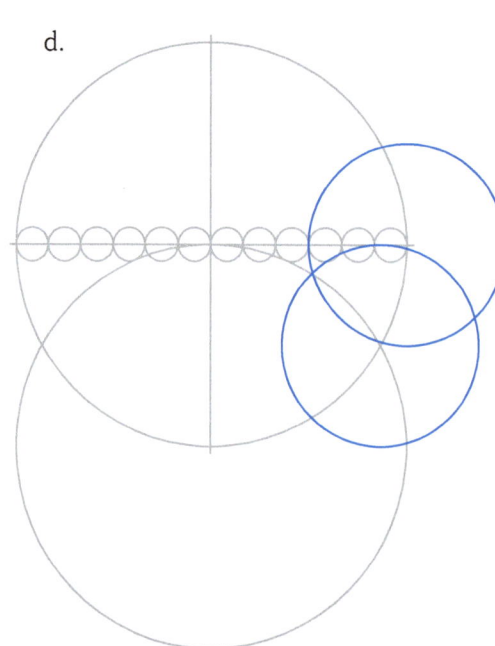

d.

a. along a straight
   line measure
   out twelve
   consecutive
   base circles

b. at center
   draw
   the base
   twelve
   circle

c. at
   quadrant
   bottom
   center a
   duplicate
   circle

d. center
   a circle
   at the newly
   created quadrant
   IV intersect and
   a duplicate at
   quadrant right

Step
Ordering

Later

Earlier

e. create a straight
   line from base 12
   origin through
   quadrant IV to
   farthest intersect
   of the two last
   circles

f. from quadrant left
   draw a straight
   line through the
   last intersect of
   quadrant IV

g. measure from
   quadrant right to
   step -f- intersect

h. draw a line from
   quadrant left
   through step
   -g- intersect of
   quadrant I

The intersects of
the Base 12 Line
will be utilized to
isolate the side
measure for each
associated circle.

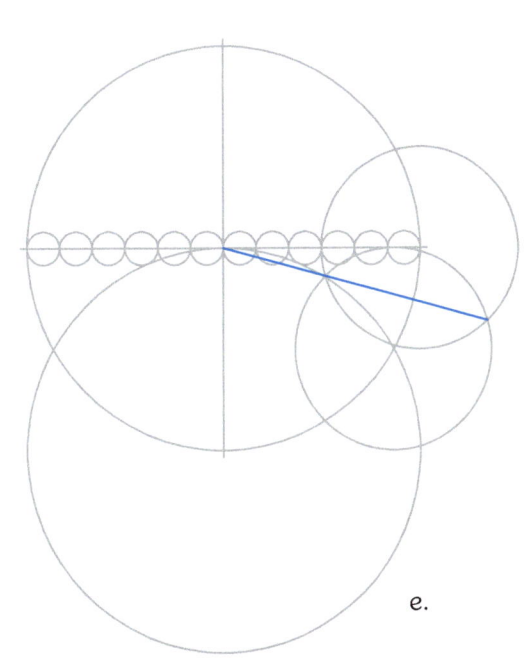

e.

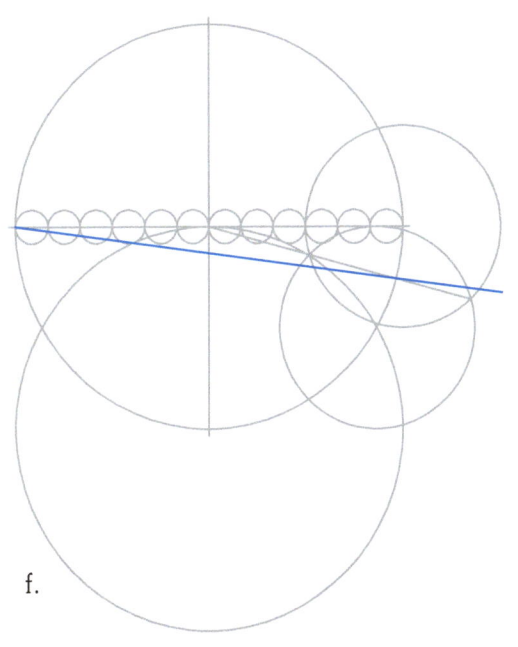

f.

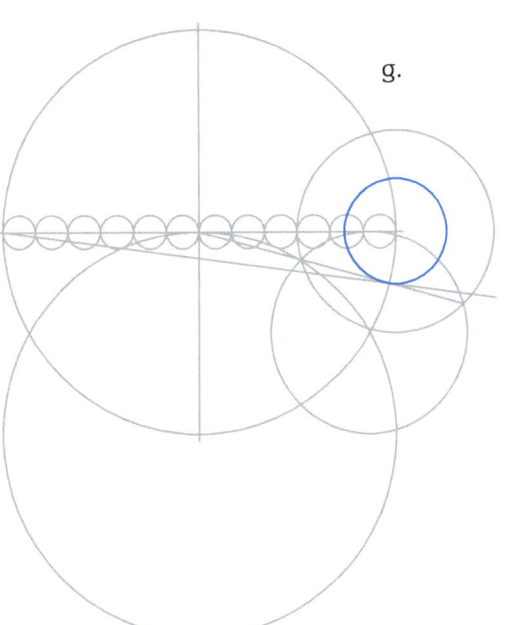

g.

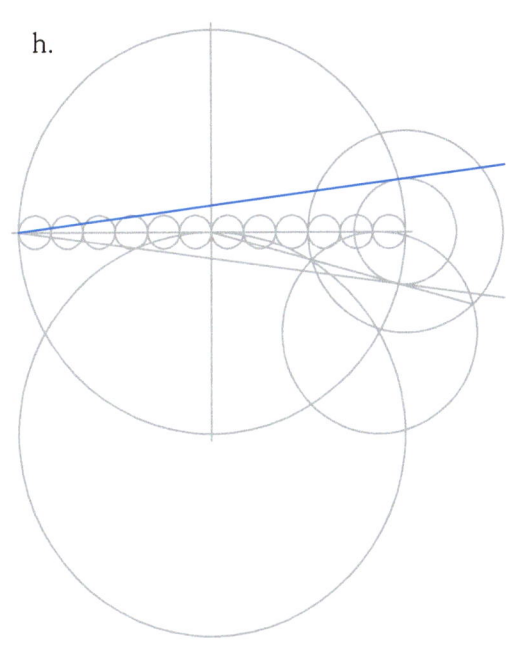

h.

eXactagon Align Method

11 Sided B12 Line ~ Hendecagon

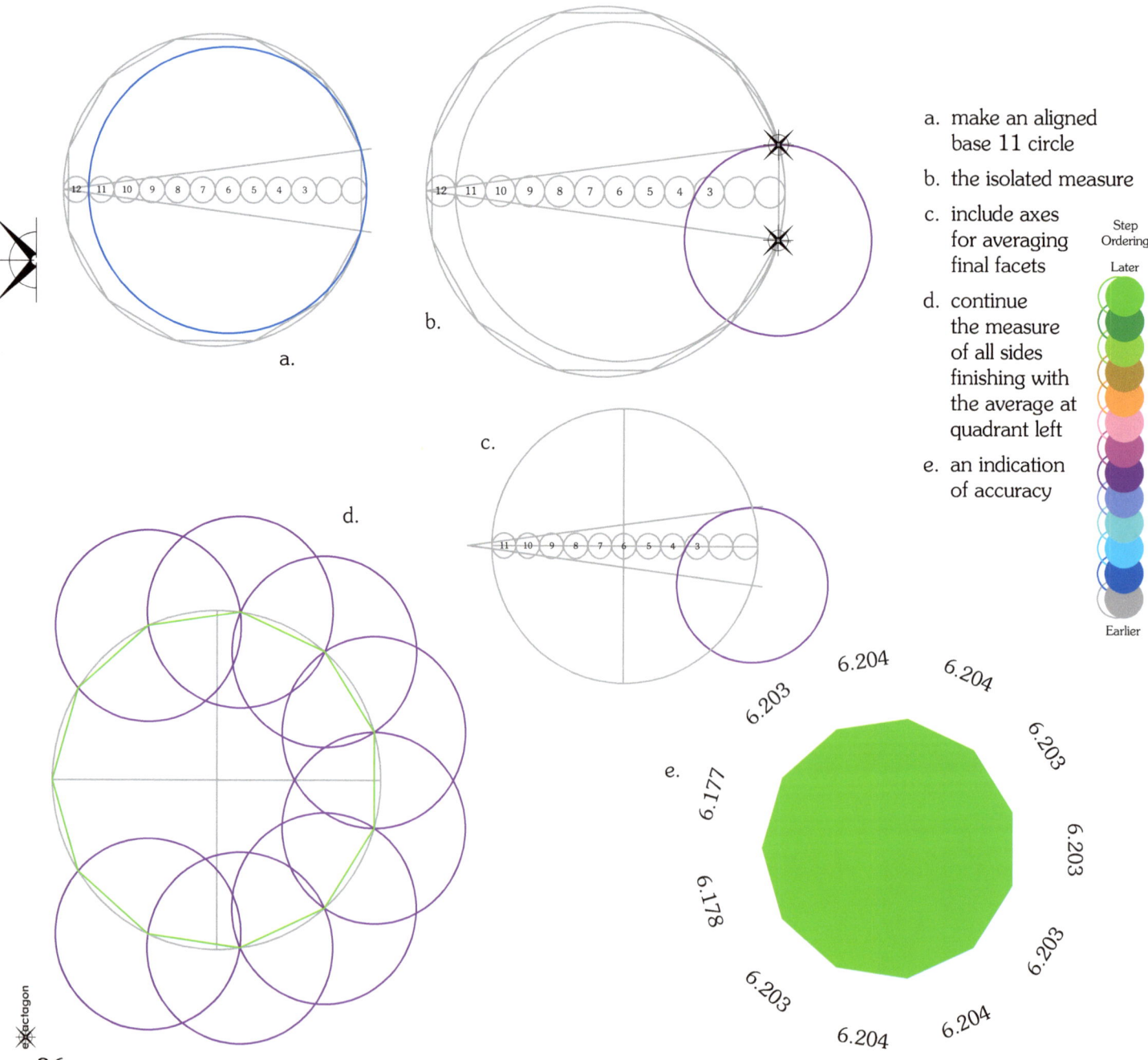

a.

b.

c.

d.

e.

a. make an aligned base 11 circle

b. the isolated measure

c. include axes for averaging final facets

d. continue the measure of all sides finishing with the average at quadrant left

e. an indication of accuracy

Step Ordering

Later

Earlier

6.204 6.204 6.203
6.203 6.203
6.177
6.203
6.178
6.203
6.204 6.204
6.203

eXactagon

13 Sided B12 Line ~ Tridecagon

a. make an aligned
base 13 circle

b. the isolated measure

c. include axes for
averaging final facets

d. continue the
measure of all sides
finishing with the
average at
quadrant left

e. an indication of
accuracy

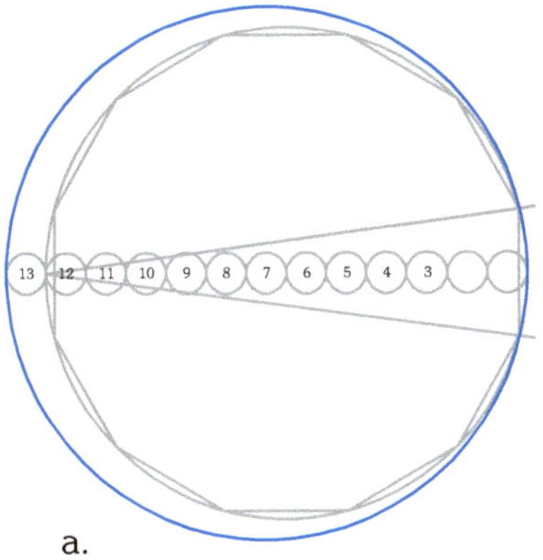

a.

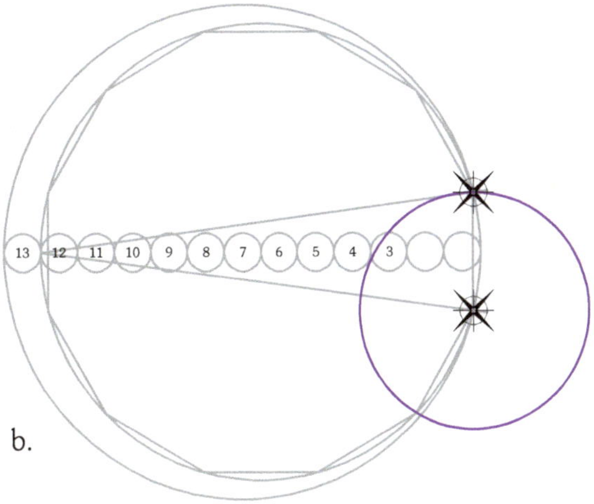

b.

c.

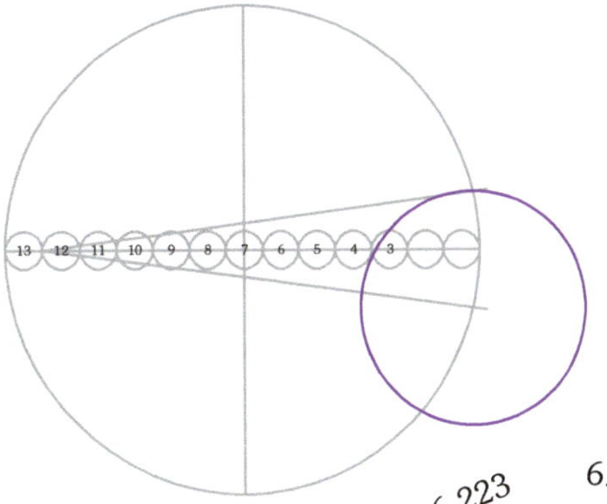

d.

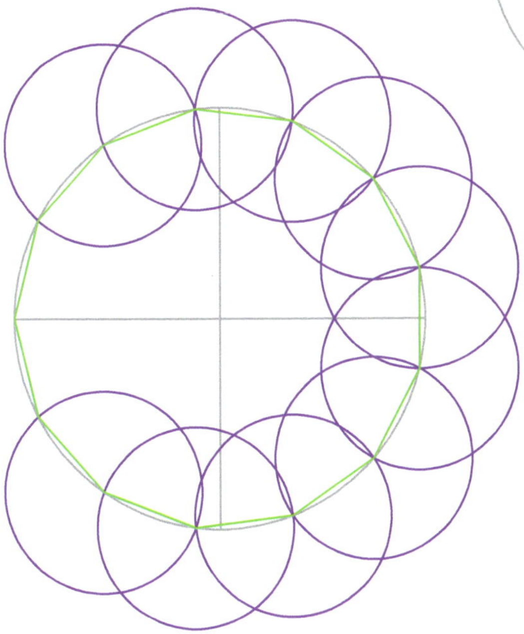

e.

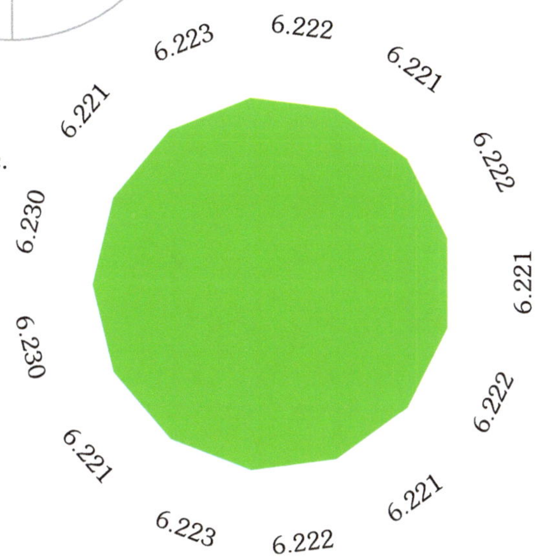

6.223 6.222
6.221   6.221
6.222
6.230   6.221
6.030
6.221
6.230   6.222
6.221   6.221
6.223 6.222

Align Method :: 11 Sided B12 Line ~ Hendecagon · 13 Sided B12 Line ~ Tridecagon

eXactagon Align Method

15 Sided B12 Line ~ Pentadecagon

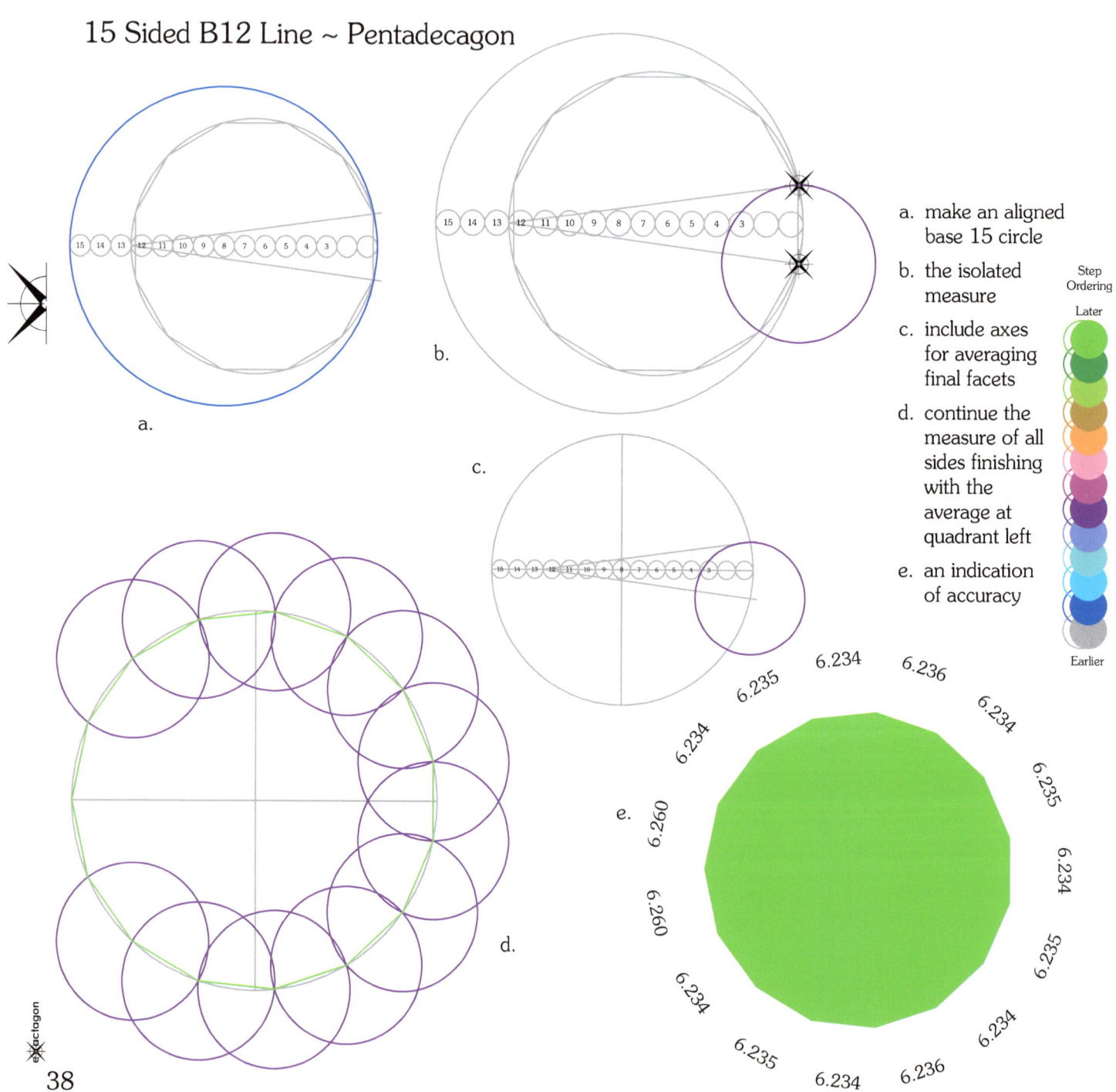

a. make an aligned base 15 circle

b. the isolated measure

c. include axes for averaging final facets

d. continue the measure of all sides finishing with the average at quadrant left

e. an indication of accuracy

Step Ordering

Later

Earlier

17 Sided B12 Line ~ Heptadecagon

a. make an aligned base 17 circle

b. the isolated measure

c. include axes for averaging final facets

d. continue the measure of all sides finishing with the average at quadrant left

e. an indication of accuracy

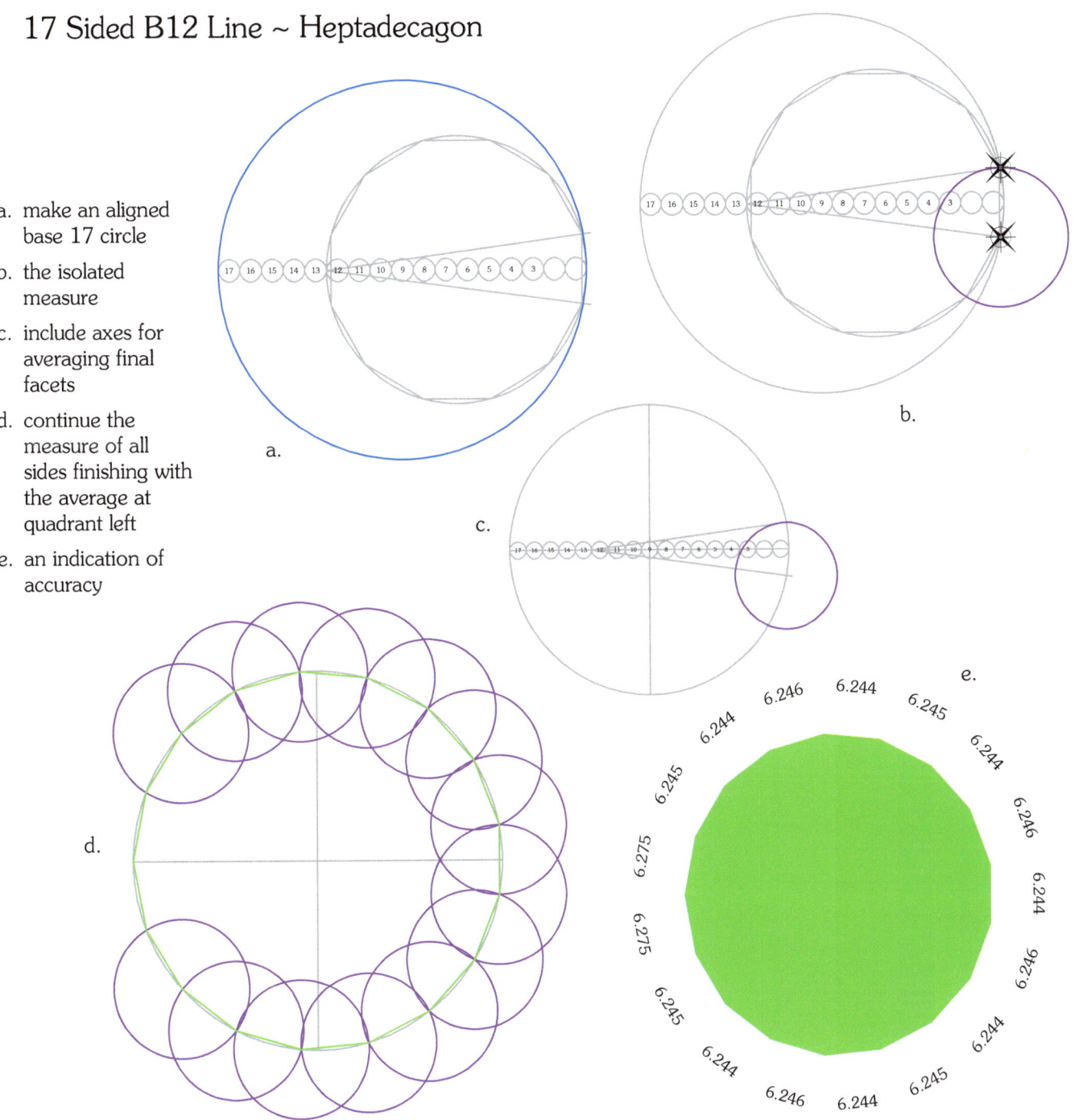

a.

b.

c.

d.

e.

✳ Align Method :: 15 Sided B12 Line ~ Pentadecagon · 17 Sided B12 Line ~ Heptadecagon

39

eXactagon Align Method

18 Sided B12 Line ~ Octadecagon

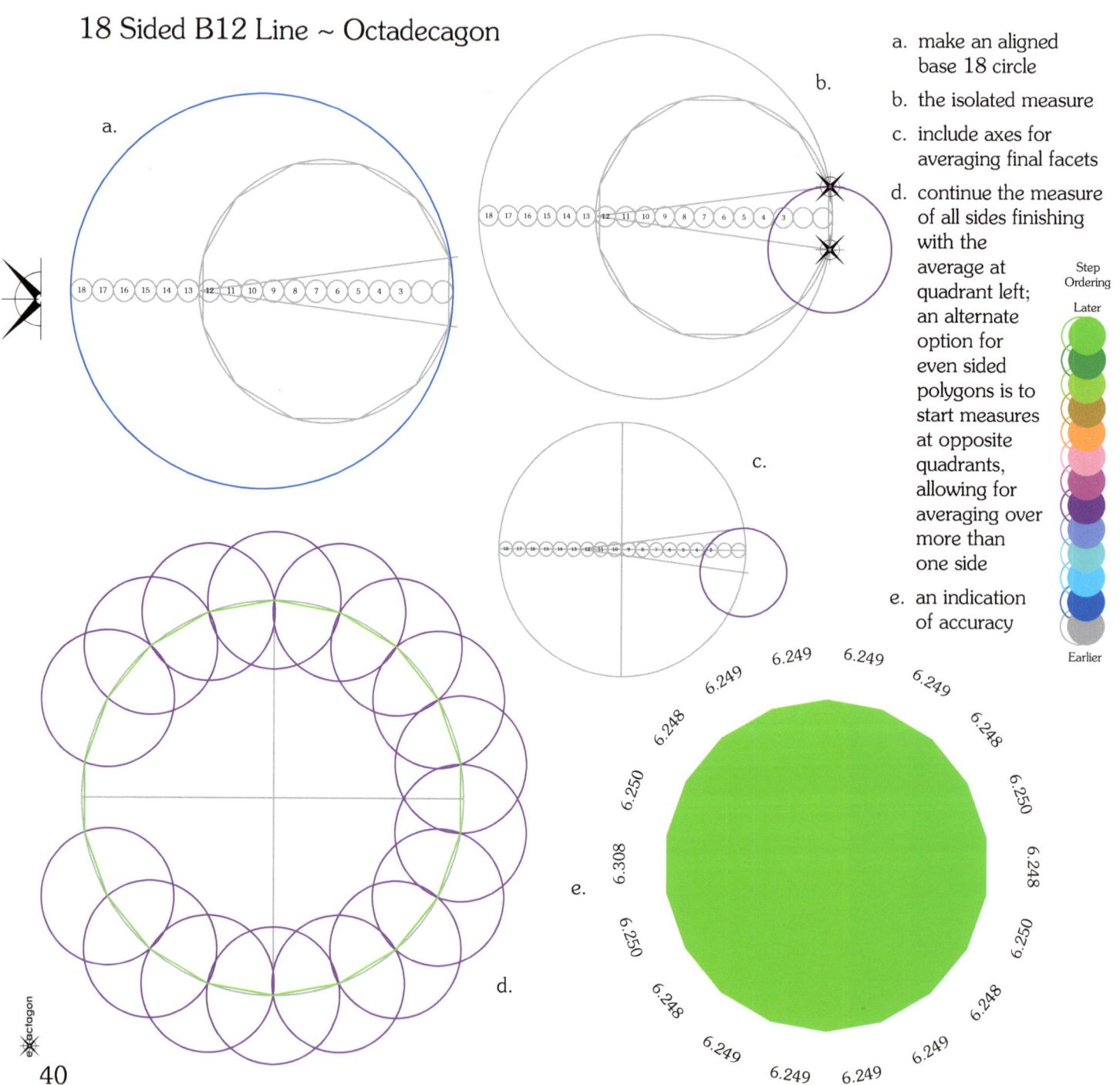

a. make an aligned base 18 circle

b. the isolated measure

c. include axes for averaging final facets

d. continue the measure of all sides finishing with the average at quadrant left; an alternate option for even sided polygons is to start measures at opposite quadrants, allowing for averaging over more than one side

e. an indication of accuracy

Step Ordering

Later

Earlier

19 Sided B12 Line ~ Enneadecagon

a. make an aligned base 19 circle

b. the isolated measure

c. include axes for averaging final facets

d. continue the measure of all sides finishing with the average at quadrant left

e. an indication of accuracy

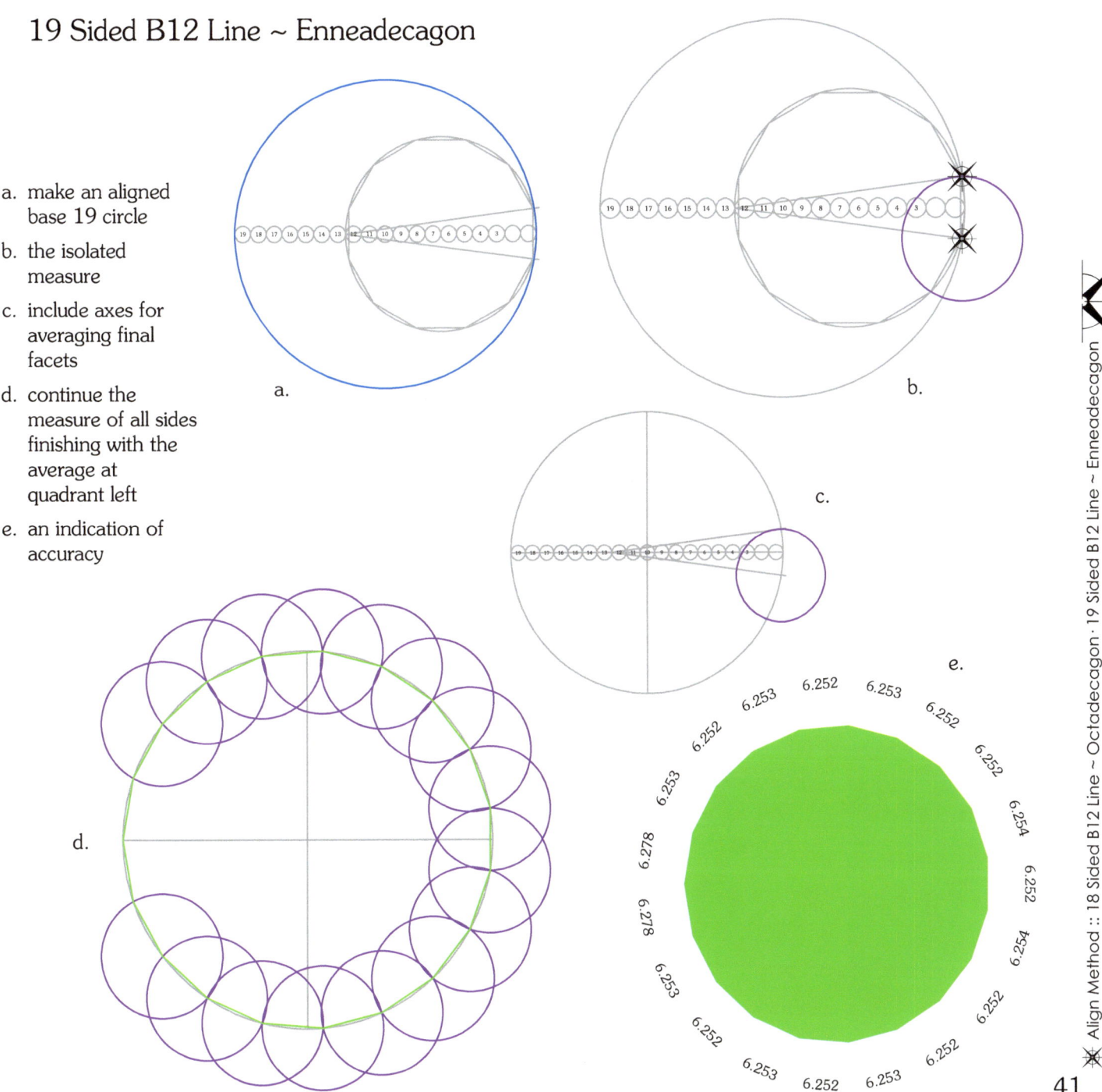

a.

b.

c.

d.

e.

Align Method :: 18 Sided B12 Line ~ Octadecagon · 19 Sided B12 Line ~ Enneadecagon

6.252 6.253 6.252 6.253 6.252 6.254 6.252 6.254 6.252 6.253 6.252 6.253 6.252 6.253 6.278 6.278 6.253 6.253

41

eXactagon Align Method

Color Stepped Composites

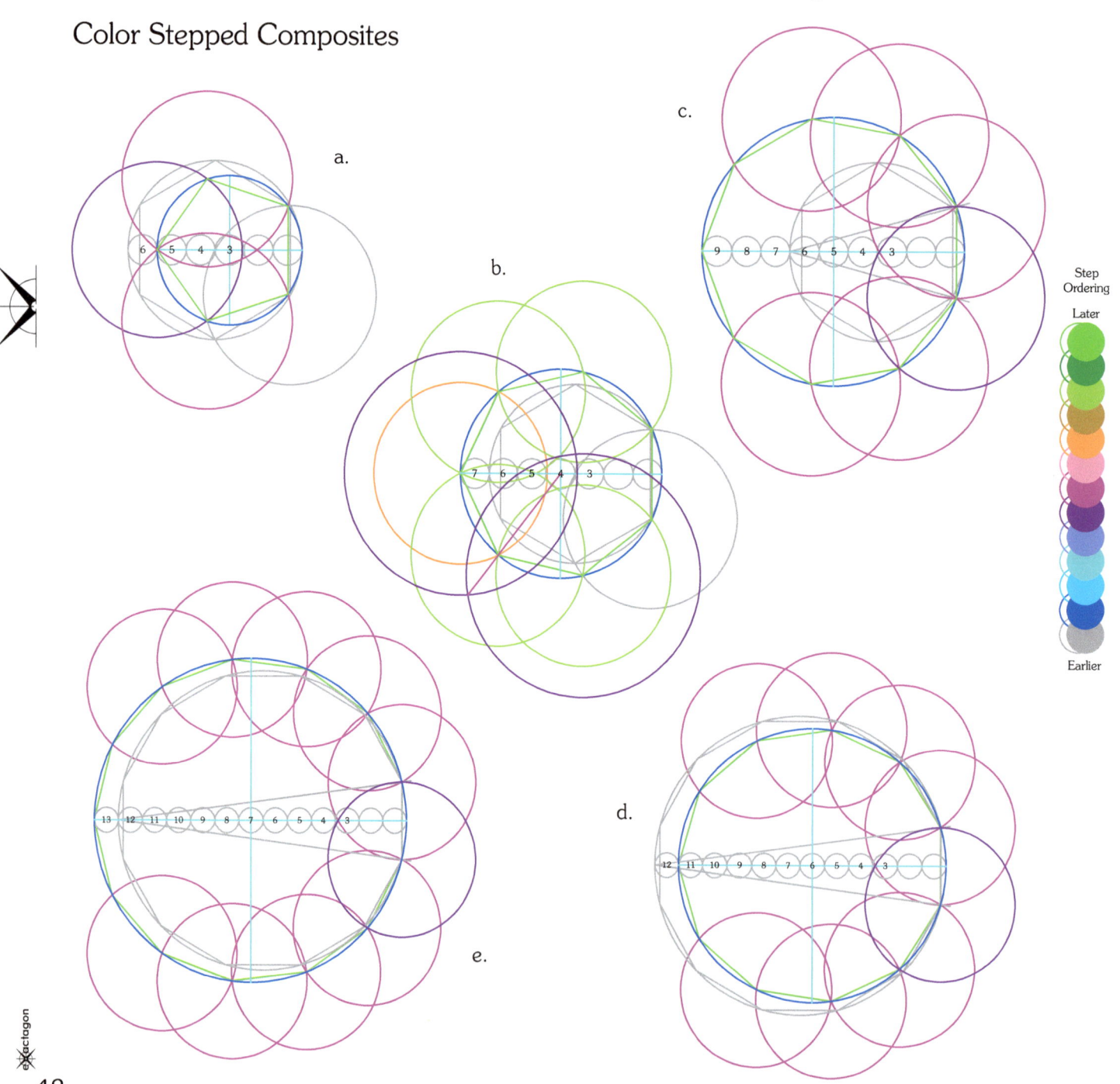

a.

b.

c.

d.

e.

Step
Ordering

Later

Earlier

a. pentagon
b. heptagon
c. enneagon
d. hendecagon
e. tridecagon
f. pentadecagon
g. heptadecagon
h. octadecagon
i. enneadecagon

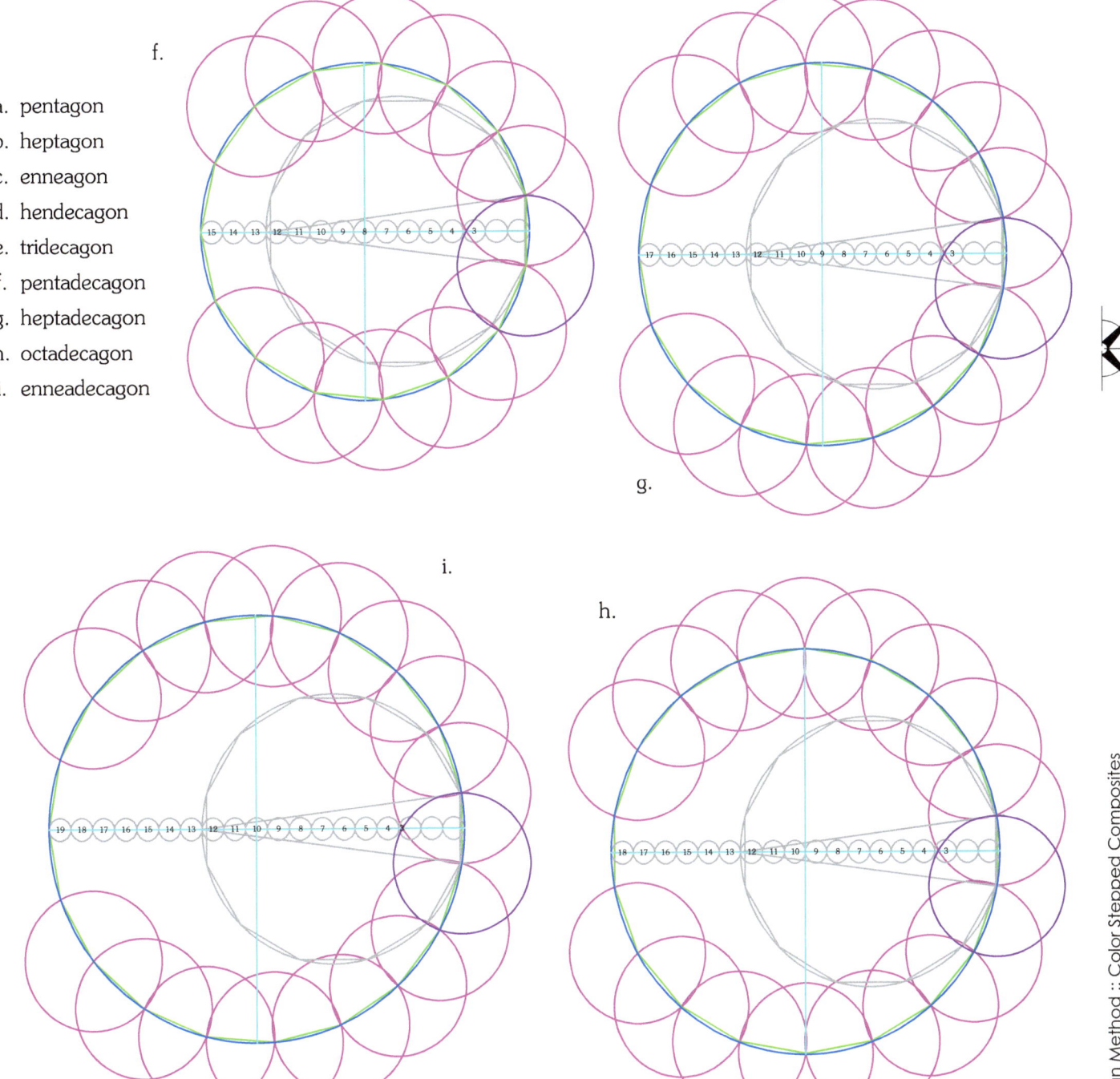

f.

g.

i.

h.

Align Method :: Color Stepped Composites

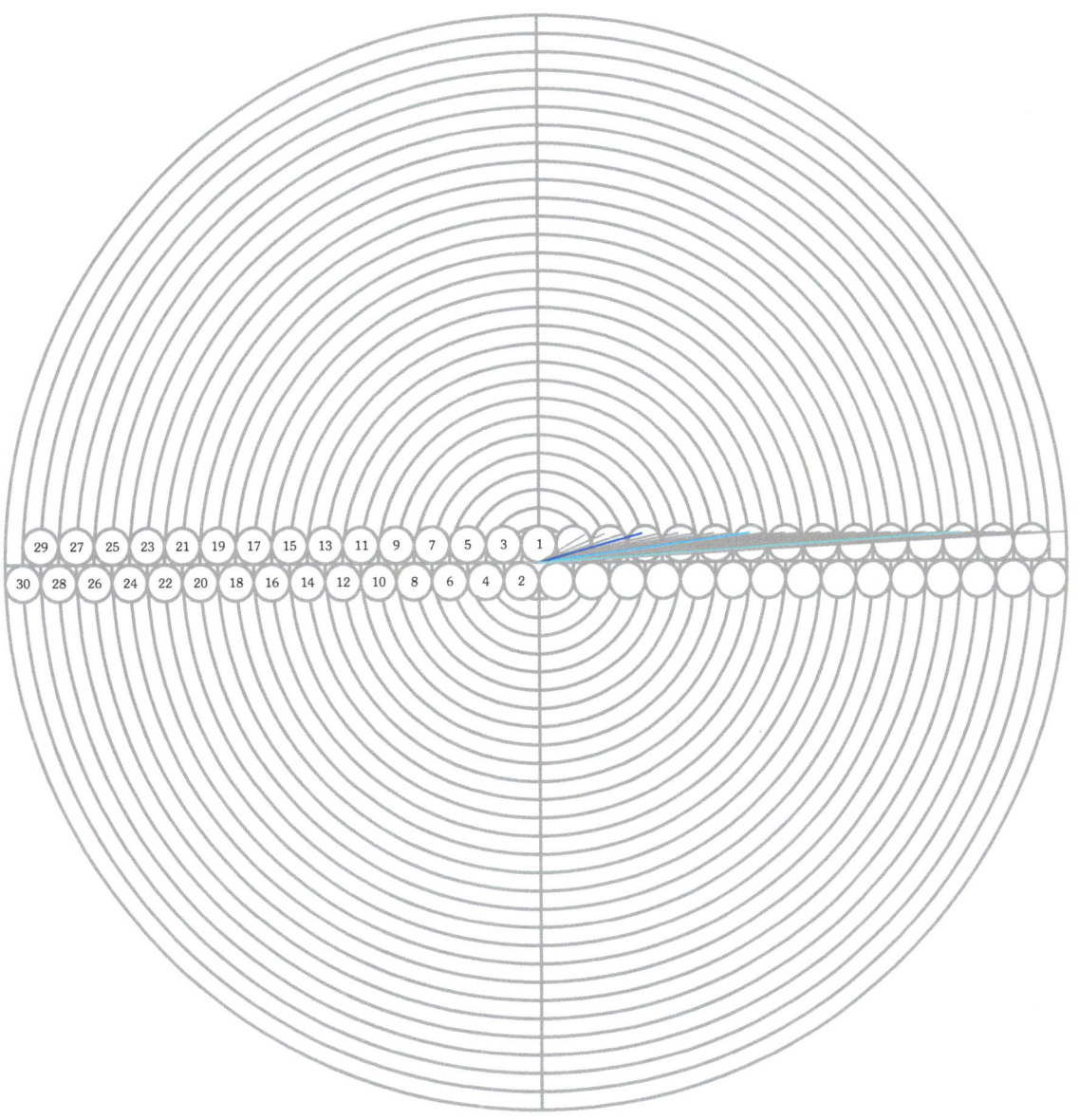

eXactagon Ratio Method

eXactagon Ratio Method

Primer

The radius for the base circle is a unit of one, and all other shown measurements and ratios are in reference to that base unit.

1.000

When reviewing the figures throughout this presentation, please be aware that slight anomalies are apparent in dimensioning and snap created precision. Such as theoretically identical side measurements not matching.

The CAD software still produced remarkable results, and any variations should easily be within the realm of human hand drawn capabilities.

*e*Xactagon Ratio Method is based on the fact that a circle can be precisely divided into four equal quadrants, which can evolve into the notion that if one quadrant can be easily divided into any number of equal segments, four of those segments should equal the side of a regular polygon of the number equaling the total segments making up the divided quadrant.

The original goal was to discover a method of dividing any angle into any number of equal sub angles. Although there may be a way that has eluded my consciousness thus far, the hexagon lends itself beautifully to the four segment notion of a single quadrant.

The quadrant of a hexagon would need to be divided into six equal parts, which is not straight forward. Yet a single side of a hexagon can be divided equally into four portions, resulting in the exact desired outcome of dividing the relevant quadrant into six equal portions.

The Ratio Method works when any other desired regular polygon has its structural circle associated with the structural circle of the hexagon. For instance a pentagon will work with the ratio measure of the hexagon if the circle housing the pentagon is five sixths the diameter of the circle for the hexagon. As well a heptagon's circle would need to be one sixth larger than the base for the hexagon.

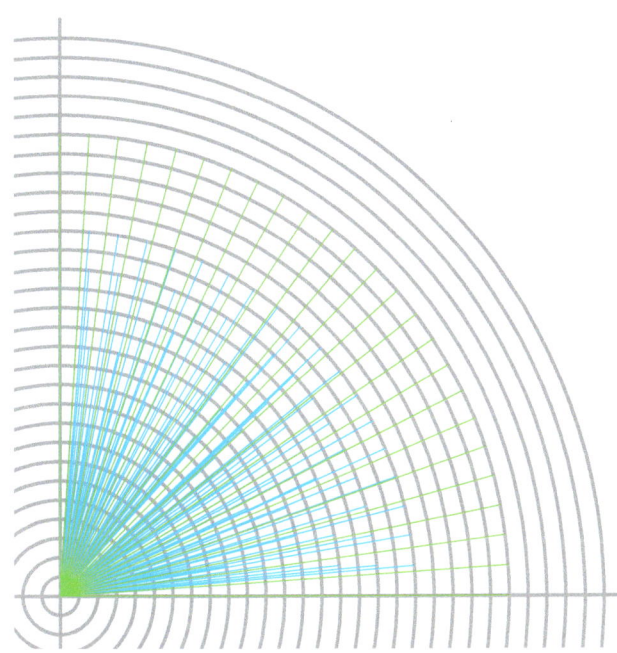

To test the accuracy of compass measure with the base six ratio, simply see if exactly four measures will span the arc of the originally divided side. Also the compass measure should total six for the span of one quadrant of the base six circle.

Further verification of accuracy is to test if ratio measure is accurate for target polygon by placing within a quadrant of associated circle the number of divisions that would equal the number of facets. The ends of the quadrant should be intersected.

For polygons of higher facets utilizing the closest stepped base polygon may be more accurate than simply relying on the hexagon or base six. All stepped bases are formed in relation to the original base six. Revealing that the next base step is a dodecagon or base twelve, then base twenty four, then forty eight and so on. Given this developing pattern each base needs to have the associated equilateral triangle arc easily divisible down to a single unit ratio. The easiest way with only a compass and straight edge is to be able to bisect each arc down to the base measure.

For the target pentagon base circle five measures of the ratio should closely fit within one quadrant. And as shown for a heptagon seven ratios should closely fit within one quadrant of the base seven circle.

Base 6

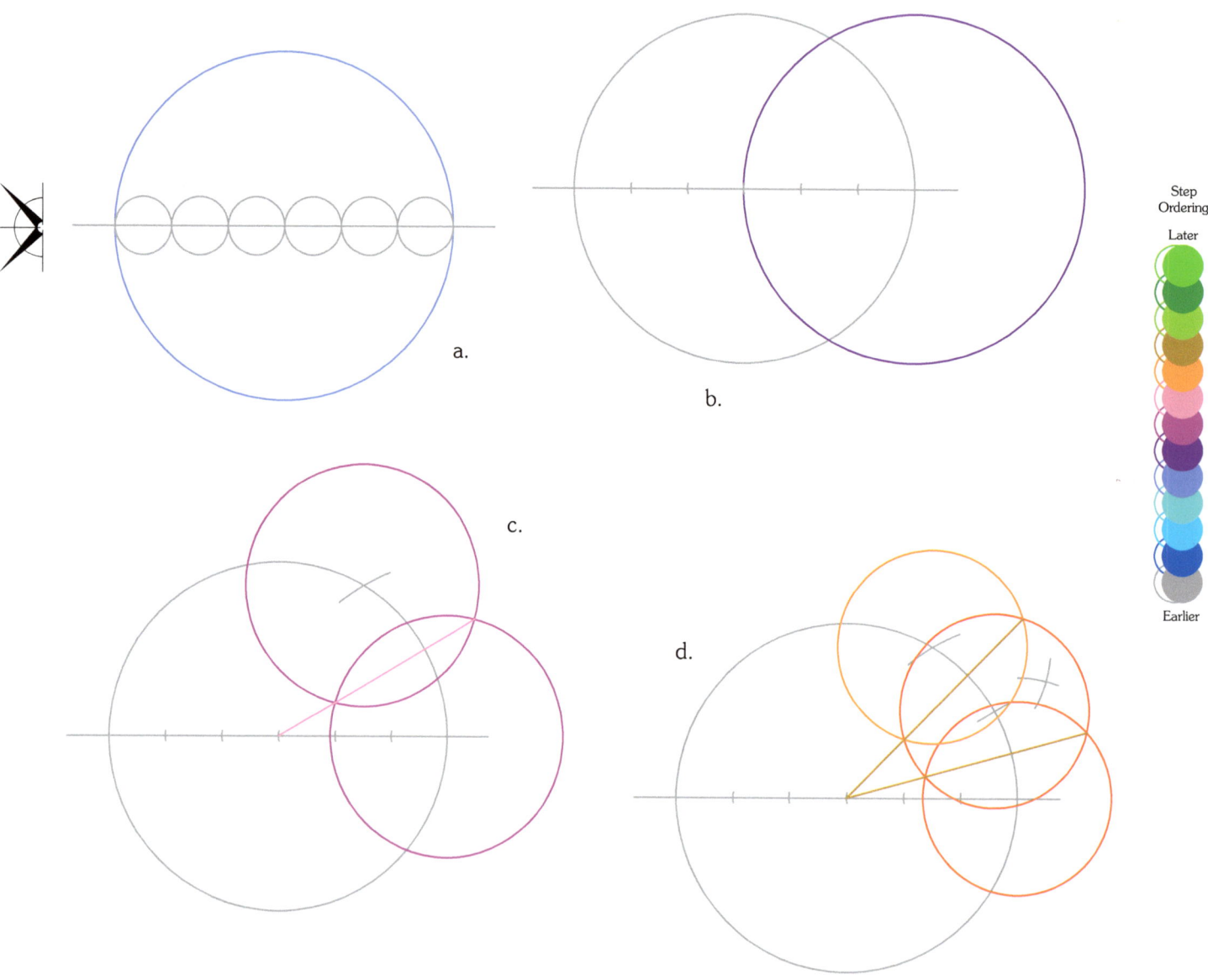

a.

b.

c.

d.

Step
Ordering

Later

Earlier

a. on a horizontal line using desired base circle, draw out six consecutive measures, then draw the resulting base six circle

b. center a duplicate circle at quadrant right

c. bisect the resulting arc associated with quadrant I

d. further bisect at least one of the resulting two arcs

e. set ratio measure to match a quarter arc section

f. verify accuracy by performing four consecutive measures to match the original quadrant I intersect of step -b-

g. ratio of base six

h. quadrant divided into six ratios

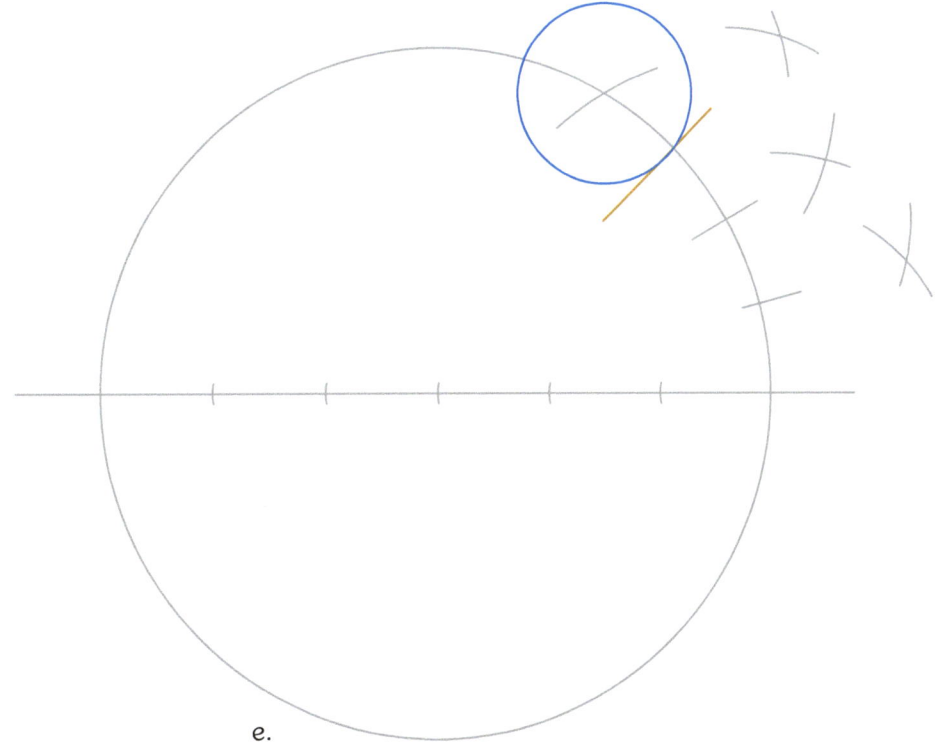

e.

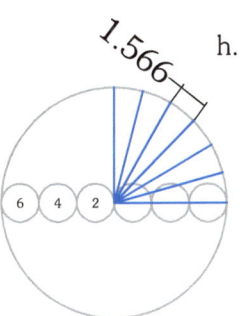

1.566

h.

g.

1.566

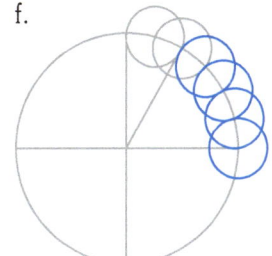

f.

eXactagon Ratio Method

5 Sided B6 ~ Pentagon

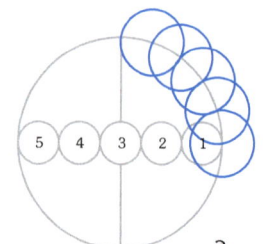

a.

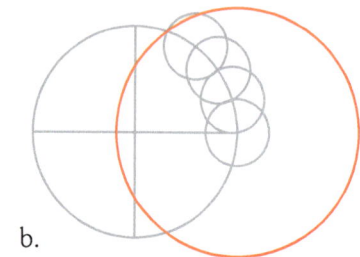

b.

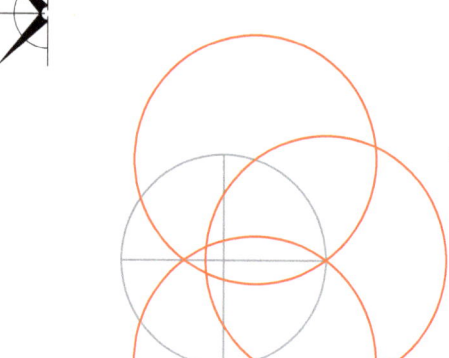

c.

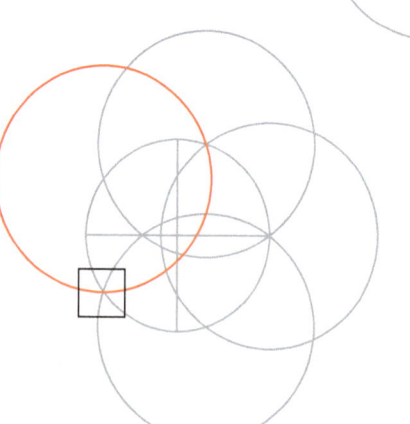

d.

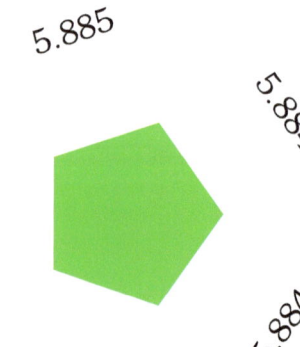

e.

f.

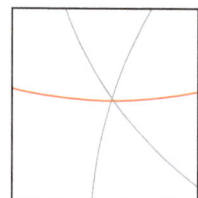

a. test measure of base 6 ratio within pentagon quadrant I

b. set compass to four ratio measures, creating a circle with radius measure of one side

c. center duplicate measures at the newly created intersects in quadrants I and IV

d. draw a line from quadrant right to each intersect in quadrants I, II, III, and IV

e. an additional side measure at quadrant II or quadrant III intersect reveals

f. discrepancy

g. side dimensions

Step Ordering

Later

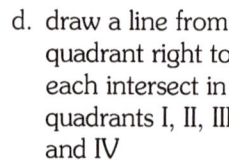
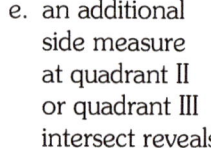

Earlier

g.

5.885

5.884

5.853

5.884

5.885

7 Sided B6 ~ Heptagon

a. test measure of base 6 ratio within heptagon quadrant I

b. set compass to four ratio measures, creating a circle with radius measure of one side

c. center duplicate measures at the newly created intersects, drawing a line from quadrant right to each intersect in quadrants I, II, III, and IV

d. an additional side measure at quadrant II or quadrant III intersect reveals

e. discrepancy

f. side dimensions

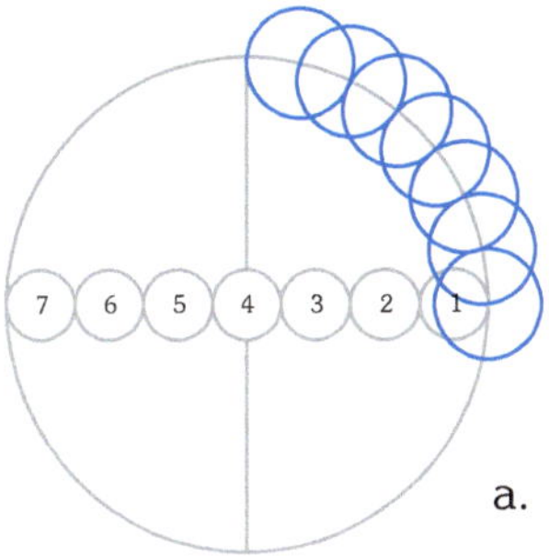

a.

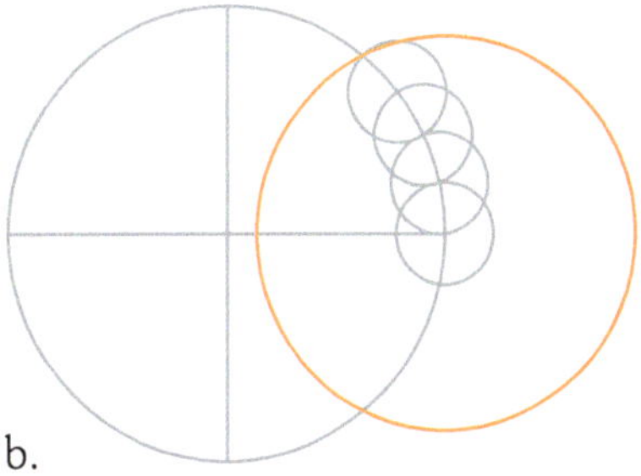

b.

c.

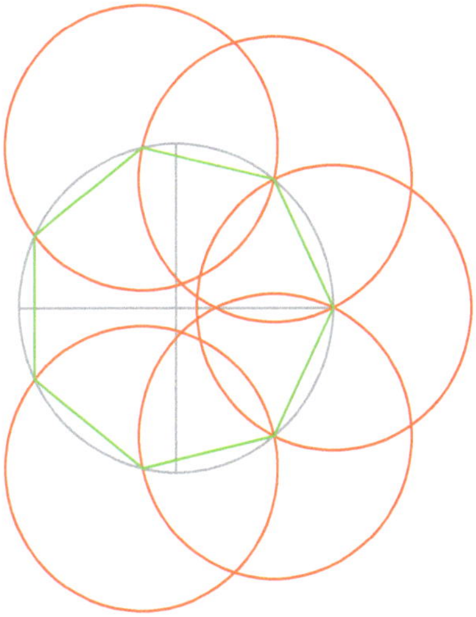

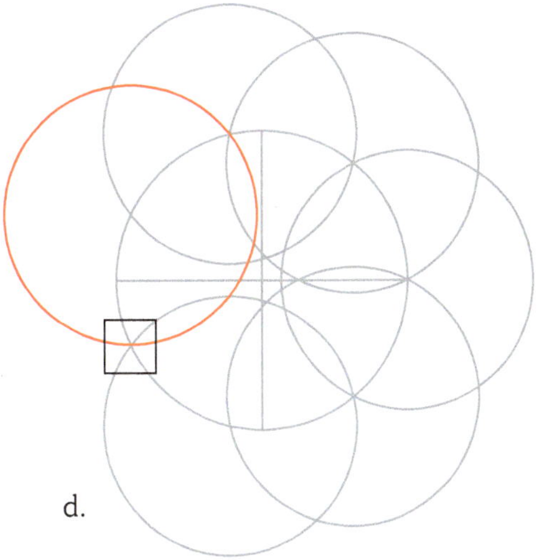

d.

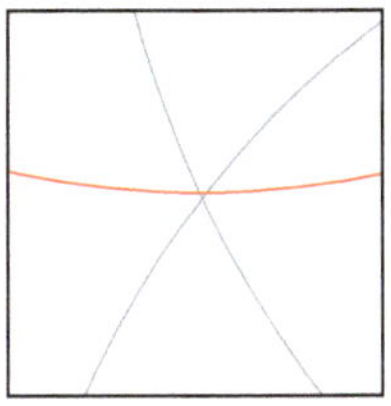

e.

f.

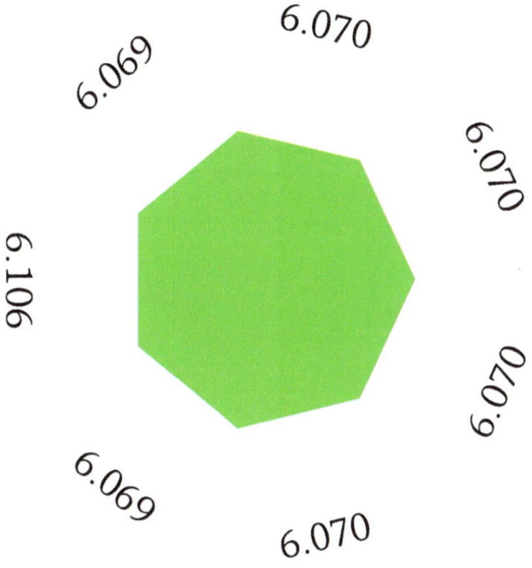

6.069

6.070

6.070

6.106

6.070

6.069

6.070

eXactagon Ratio Method

Base 12 ~ Dodecagon

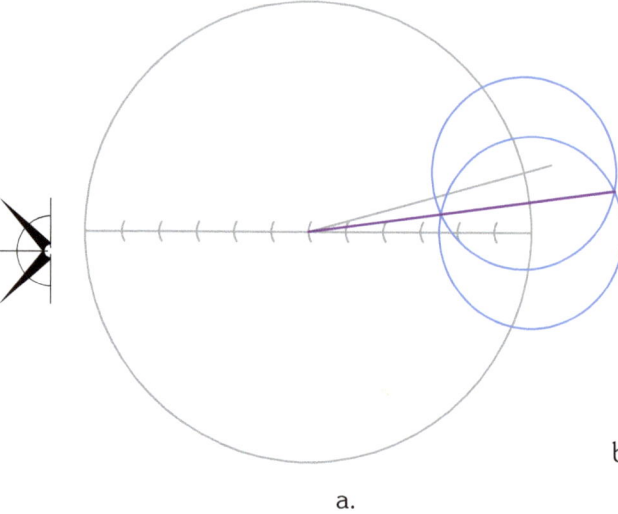

a.

a. extend line for
determining
base 6 ratio to a
base 12 circle, then
continue to bisect
the arc formed so
that the resulting
arc is an eighth
of the referenced
base 6 starting arc;
continued bisecting
will reveal consecutive
base ratios as listed in
the table "Base Steps"
on page 59 up to
base 6144

b. set compass to
measure the ratio of
base 12

c. quadrant divided into
twelve ratios

d. accuracy test

e. base 12 ratio

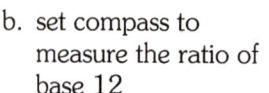

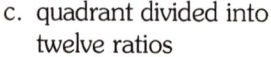

b.

Step
Ordering

Later

Earlier

1.570

c.

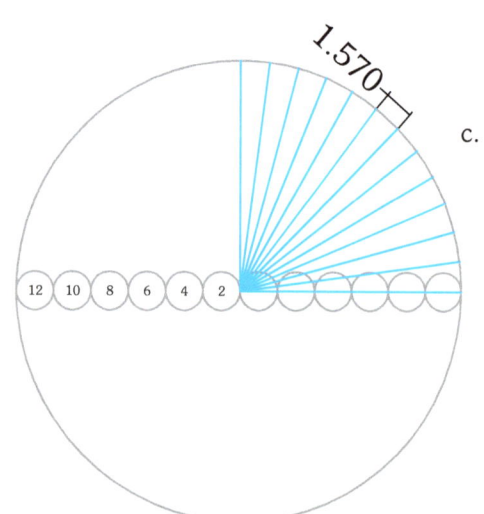

d.

1.570

e.

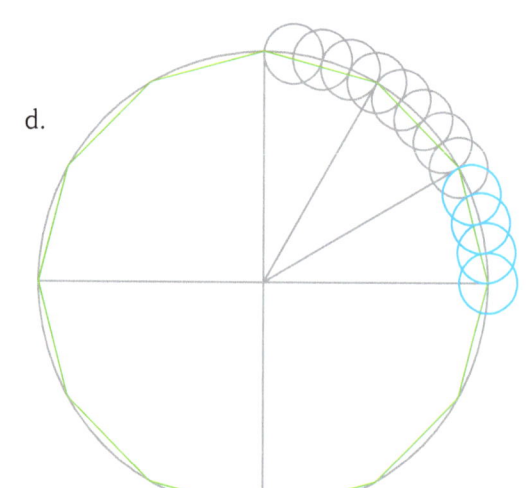

9 Sided B12 ~ Enneagon

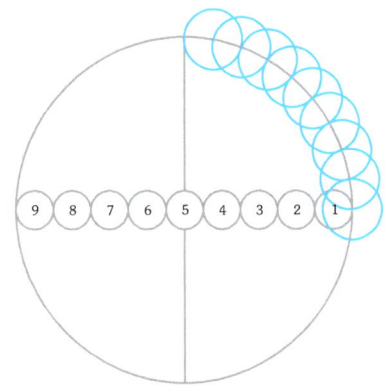

a.

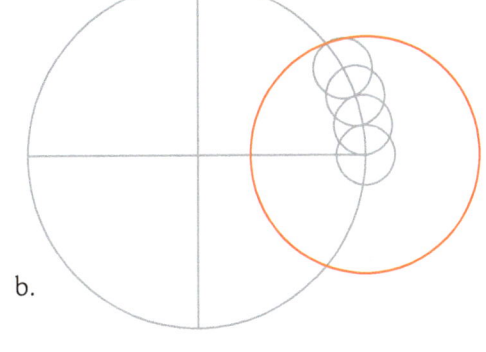

b.

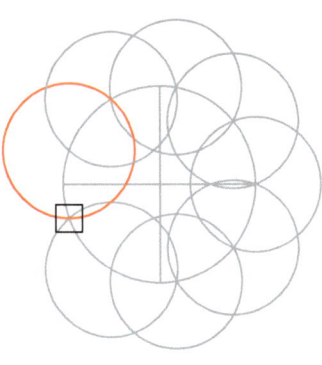

d.

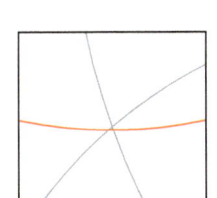

e.

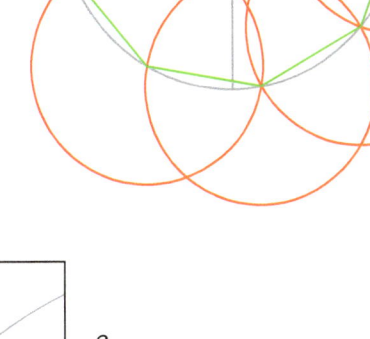

c.

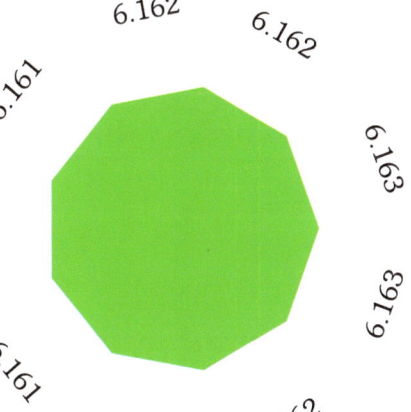

f.

a. test measure base 12 ratio within enneagon quadrant I

b. set compass to four ratio measures, creating a circle with radius measure of one side

c. center duplicate measures at each newly created intersect along target circle, working from quadrant right to quadrant left, then drawing a line from quadrant right to each intersect in quadrants I, II, III, and IV

d. an additional side measure at quadrant II or quadrant III intersect reveals

e. discrepancy

f. side dimensions

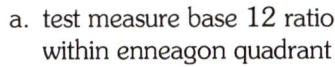
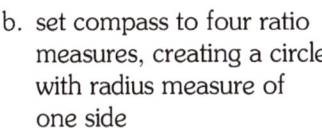
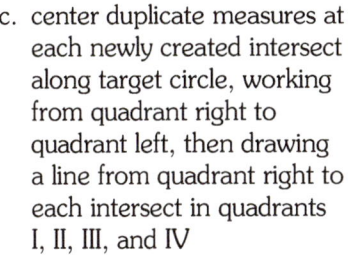
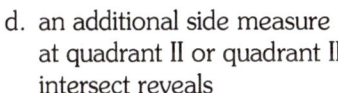

6.162 6.162
6.161 6.163
6.119 6.163
6.161 6.162
6.162

eXactagon Ratio Method

11 Sided B12 ~ Hendecagon

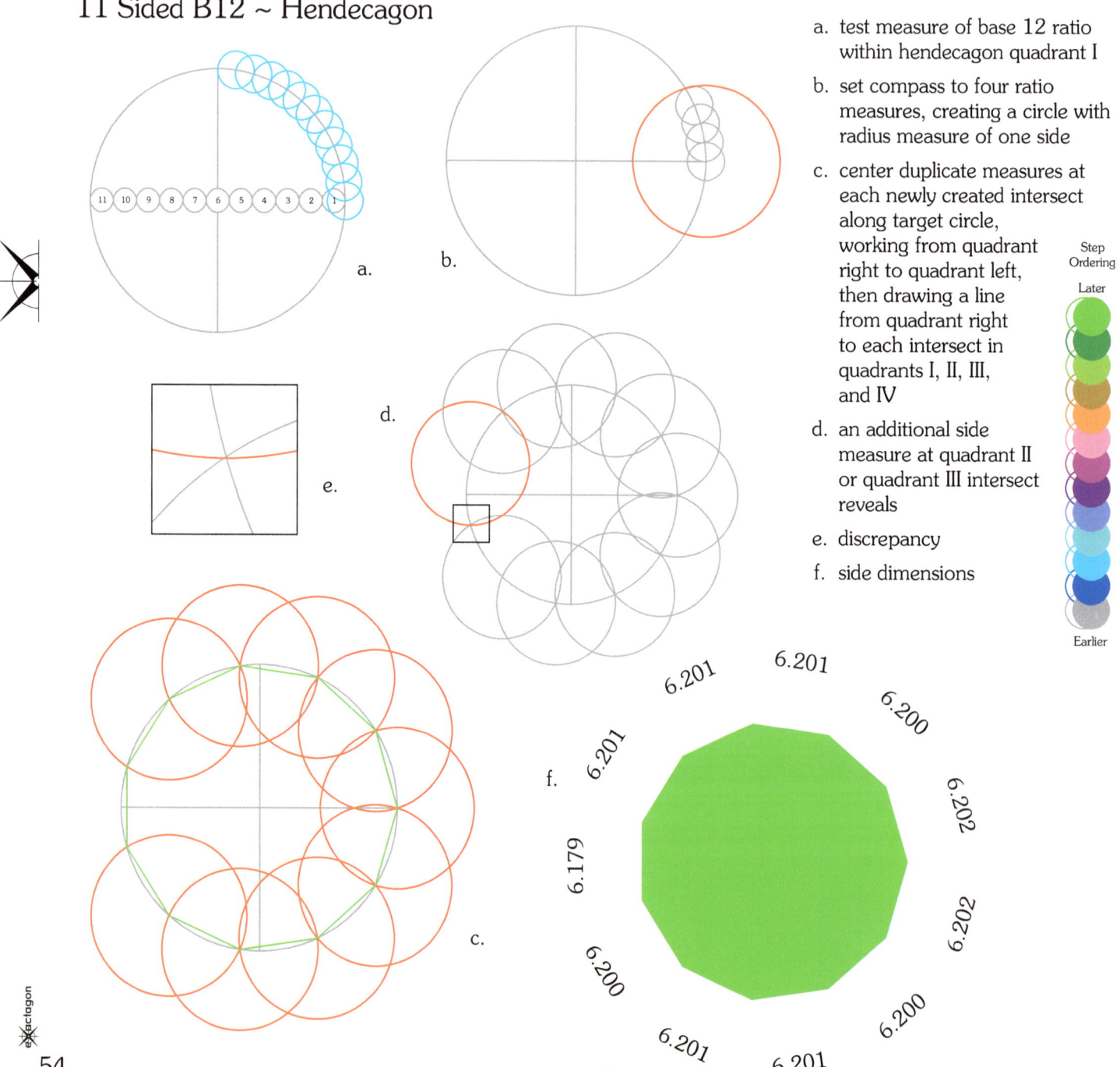

a. test measure of base 12 ratio within hendecagon quadrant I

b. set compass to four ratio measures, creating a circle with radius measure of one side

c. center duplicate measures at each newly created intersect along target circle, working from quadrant right to quadrant left, then drawing a line from quadrant right to each intersect in quadrants I, II, III, and IV

d. an additional side measure at quadrant II or quadrant III intersect reveals

e. discrepancy

f. side dimensions

Step Ordering

Later

Earlier

6.201 6.201 6.200 6.202 6.202 6.200 6.201 6.201 6.200 6.179 6.201

13 Sided B12 ~ Tridecagon

a. test measure of base 12 ratio within tridecagon quadrant I

b. set compass to four ratio measures, creating a circle with radius measure of one side

c. center duplicate measures at each newly created intersect along target circle, working from quadrant right to quadrant left, then drawing a line from quadrant right to each intersect in quadrants I, II, III, and IV

d. an additional side measure at quadrant II or quadrant III intersect reveals

e. discrepancy

f. side dimensions

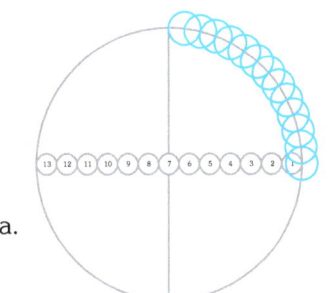

a.

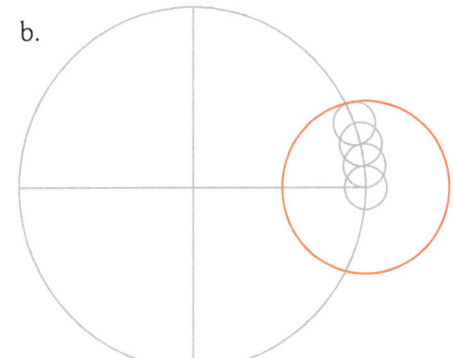

b.

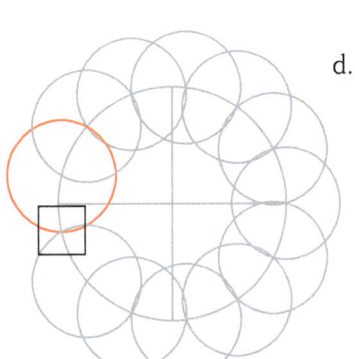

d.

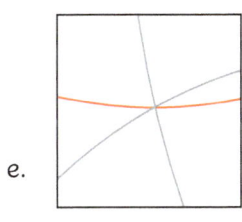

e.

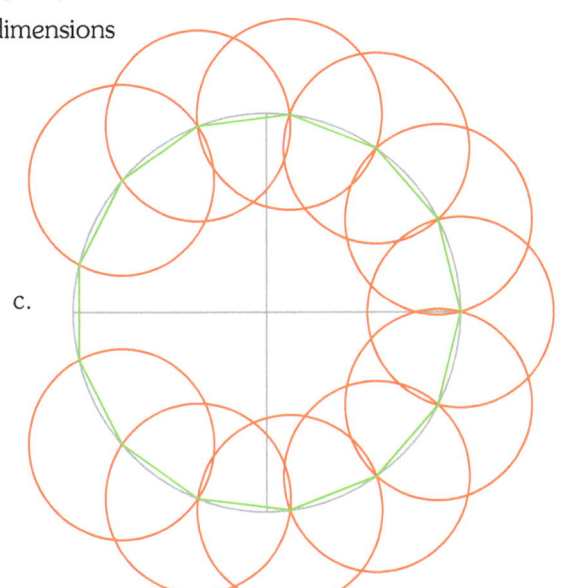

c.

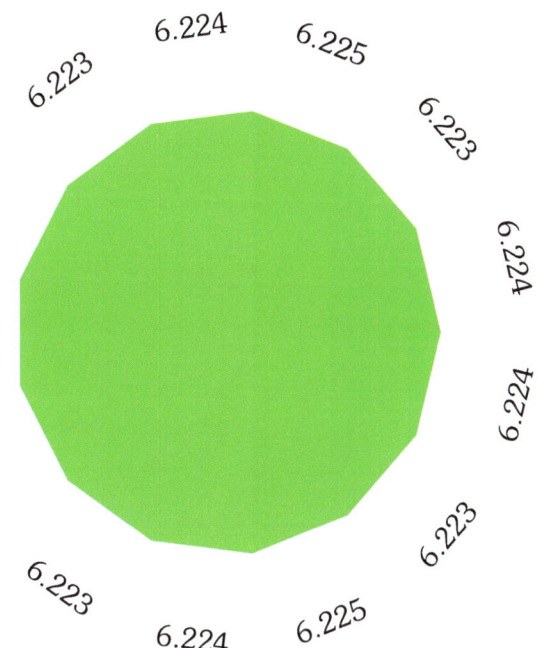

f.

6.224 · 6.225 · 6.223 · 6.223 · 6.224 · 6.224 · 6.224 · 6.223 · 6.225 · 6.224 · 6.223 · 6.224 · 6.212

eXactagon Ratio Method

15 Sided B12 ~ Pentadecagon

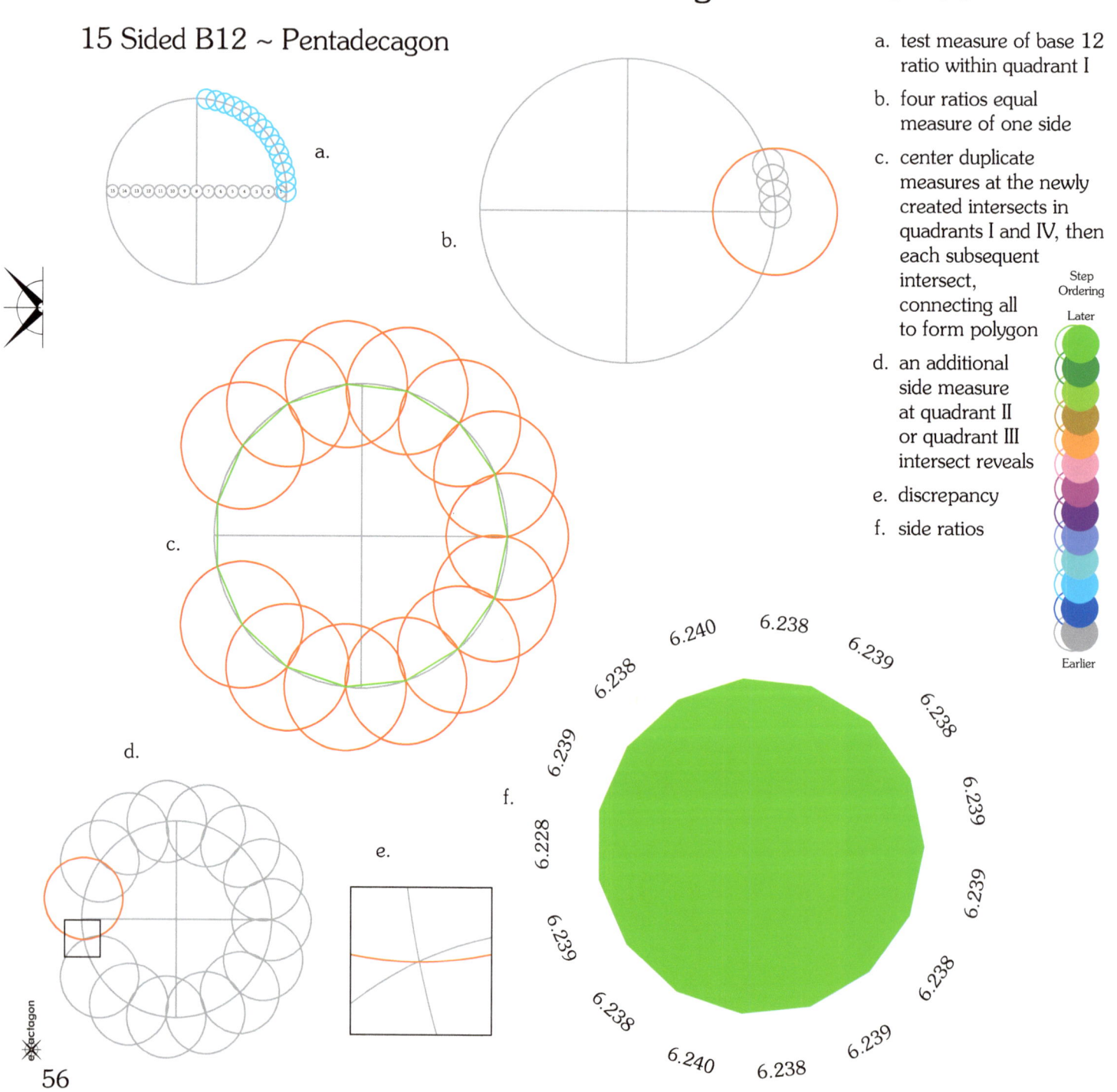

a. test measure of base 12 ratio within quadrant I

b. four ratios equal measure of one side

c. center duplicate measures at the newly created intersects in quadrants I and IV, then each subsequent intersect, connecting all to form polygon

d. an additional side measure at quadrant II or quadrant III intersect reveals

e. discrepancy

f. side ratios

Step Ordering

Later

Earlier

a.

b.

c.

d.

e.

f.

6.240 6.238 6.239
6.238   6.238
6.239   6.239
6.228   6.239
6.239   6.239
6.239   6.238
6.238   6.239
6.240 6.238

17 Sided B12 ~ Heptadecagon

a. test measure of base 12 ratio within heptadecagon quadrant I

b. set compass to four ratio measures, creating a circle with radius measure of one side

c. center duplicate measures at each newly created intersect along target circle, working from quadrant right to quadrant left, then drawing a line from quadrant right to each intersect in quadrants I, II, III, and IV

d. an additional side measure at quadrant II or quadrant III intersect reveals

e. discrepancy

f. side dimensions

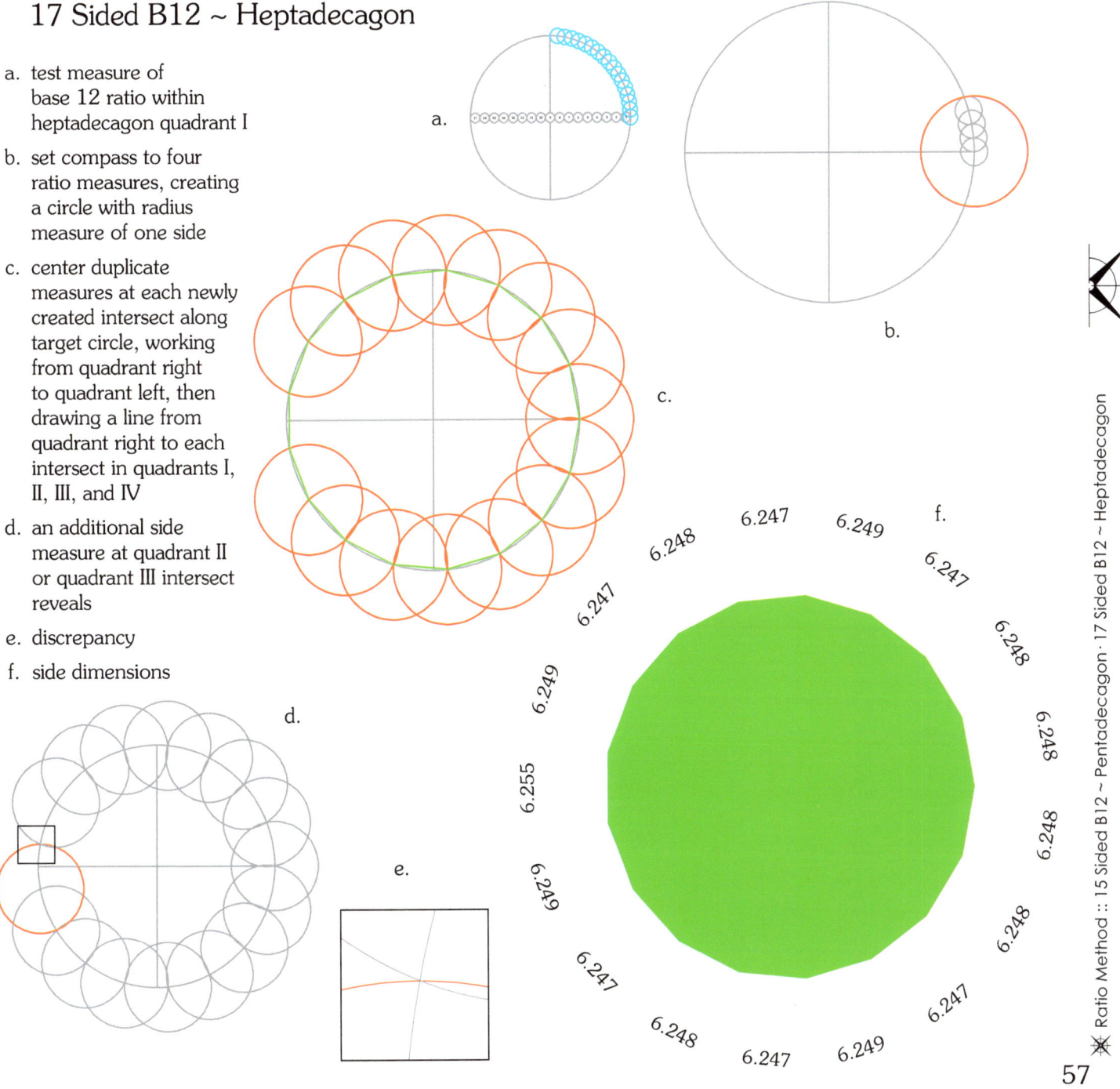

a.

b.

c.

d.

e.

f.

6.247 6.248 6.249 6.247 6.247 6.248 6.248 6.248 6.248 6.248 6.247 6.249 6.247 6.248 6.247 6.249 6.247 6.255 6.249

eXactagon Ratio Method

19 Sided B12 ~ Enneadecagon

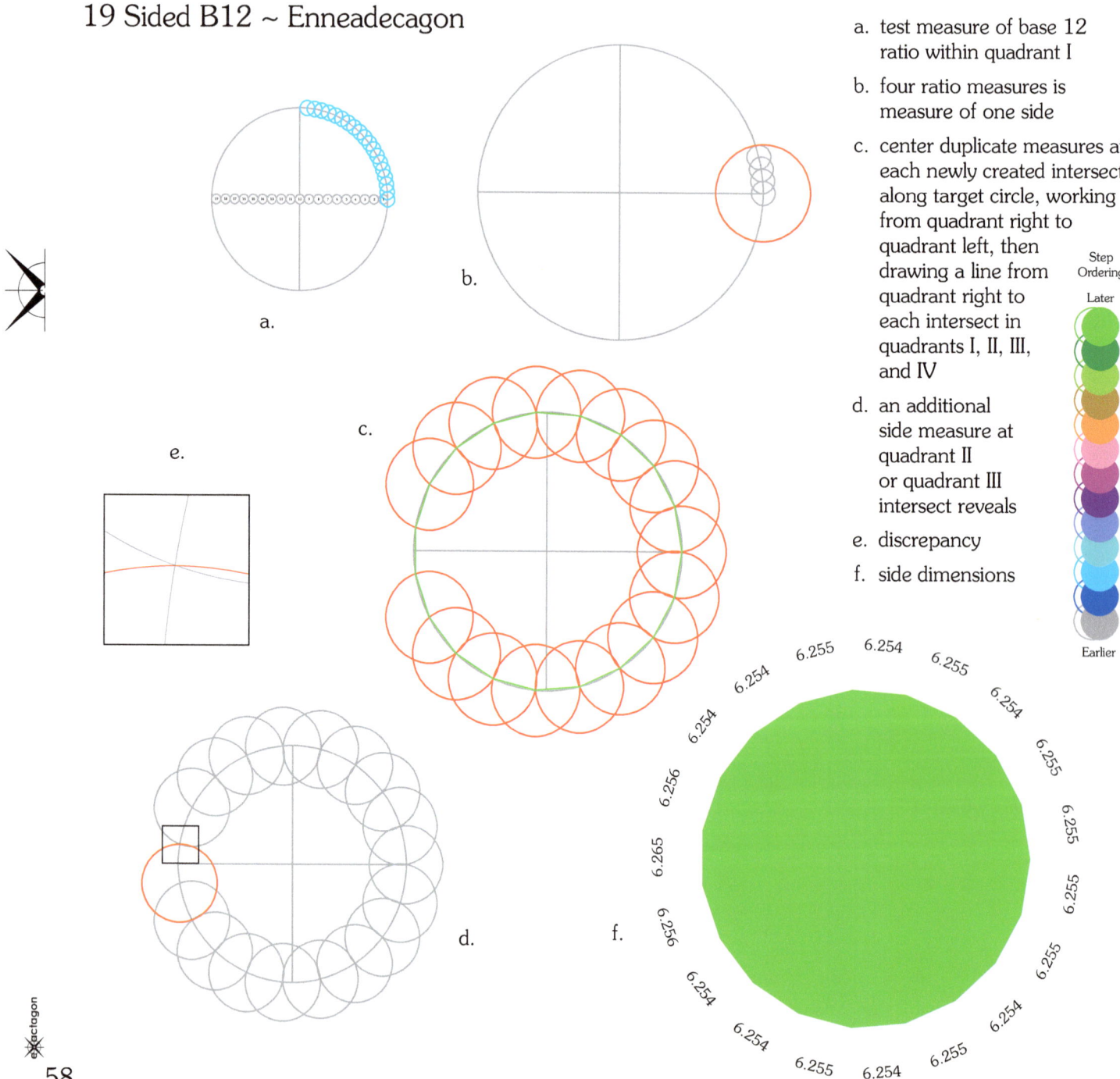

a. test measure of base 12 ratio within quadrant I

b. four ratio measures is measure of one side

c. center duplicate measures at each newly created intersect along target circle, working from quadrant right to quadrant left, then drawing a line from quadrant right to each intersect in quadrants I, II, III, and IV

d. an additional side measure at quadrant II or quadrant III intersect reveals

e. discrepancy

f. side dimensions

Step Ordering

Later

Earlier

a.

b.

c.

d.

e.

f.

6.254 6.255 6.254 6.255 6.254 6.255 6.255 6.255 6.255 6.255 6.254 6.255 6.254 6.255 6.254 6.254 6.256 6.256 6.265

Base 24 ~ Icositetragon

a. extend the base 12 line to the base 24 circle then bisect so that the resulting arc is a sixteenth of the starting base 6 arc

b. set compass to measure the ratio of base 24

c. quadrant divided into 24 ratios

d. accuracy test

e. base 24 ratio

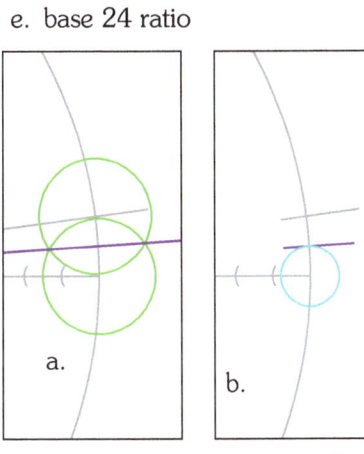

a.

b.

c.

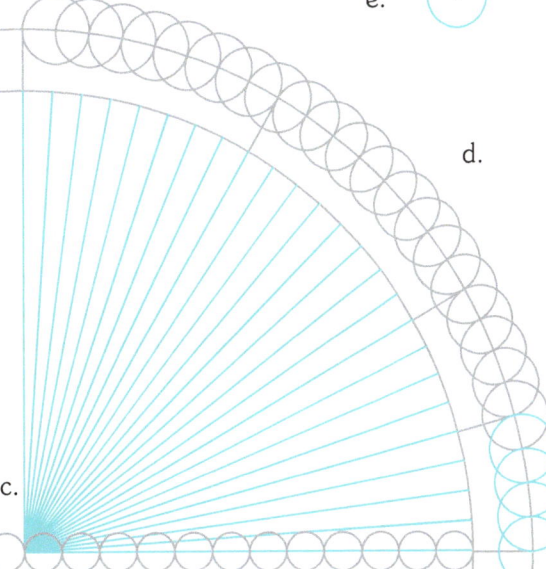

1.571

e.

d.

The following table shows single base ratio measures and double spanned ratio measures since a clear association became present.

Revealed in the table is a strong relation to pi within the Ratio Method. When associated with base setup circles, accuracy is closest at base 384, and a pinnacle is reached at around base 768. For the purpose of this eXactagon presentation, no further limits were sought.

Base Steps	Rounded	Max Measured 1.5708039709	Doubled	Max Measured 3.1416049956	Rounded	Compared with pi 3.141592654
6	1.566	1.5663167979	3.13263360	3.1058338190	3.106	-1.13823906%
12	1.570	1.5696765713	3.13935314	3.1326264385	3.133	-0.28540349%
24	1.571	1.5705147909	3.14102958	3.1393460527	3.139	-0.07151153%
48	1.571	1.5707260249	3.14145205	3.1410321404	3.141	-0.01784169%
96	1.571	1.5707769743	3.14155395	3.1414519854	3.141	-0.00447761%
192	1.571	1.5707926685	3.14158534	3.1415578855	3.142	-0.00110670%
384	1.571	1.5707946238	3.14158925	3.1415853370	3.142	-0.00023289%
768	1.571	1.5708039709	3.14160794	3.1416049956	3.142	0.00039286%
1536	1.571	1.5708033549	3.14160671	3.1416040048	3.142	0.00036132%
3072	1.571	1.5707874510	3.14157490	3.1415823480	3.142	-0.00032804%
6144	1.571	1.5707480440	3.14149609	3.1414977600	3.141	-0.00302056%

Compound Factor : 62

30 Sided B12 ~ Triacontagon

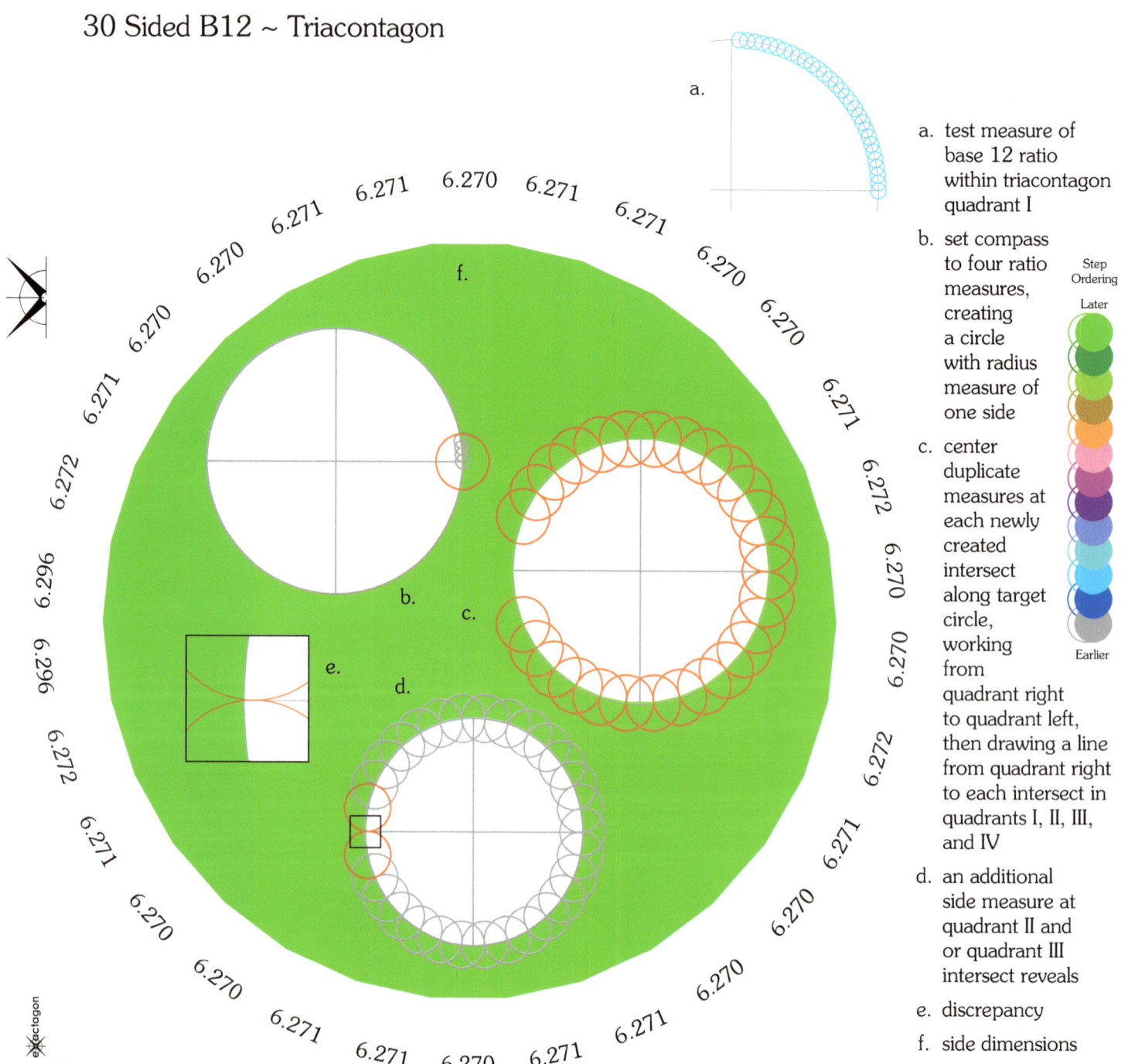

a. test measure of base 12 ratio within triacontagon quadrant I

b. set compass to four ratio measures, creating a circle with radius measure of one side

Step Ordering
Later

c. center duplicate measures at each newly created intersect along target circle, working from quadrant right to quadrant left, then drawing a line from quadrant right to each intersect in quadrants I, II, III, and IV

Earlier

d. an additional side measure at quadrant II and or quadrant III intersect reveals

e. discrepancy

f. side dimensions

30 Sided B24 ~ Triacontagon

a. test measure of base 24
   ratio within triacontagon
   quadrant I

b. set compass to four
   ratio measures, creating
   a circle with radius
   measure of one side

c. center duplicate
   measures at each newly
   created intersect along
   target circle, working
   from quadrant right
   to quadrant left,
   then drawing
   a line from
   quadrant right to
   each intersect in
   quadrants I, II,
   III, and IV

d. an additional
   side measure at
   quadrant II and
   or quadrant III
   intersect reveals

e. discrepancy

f. side dimensions

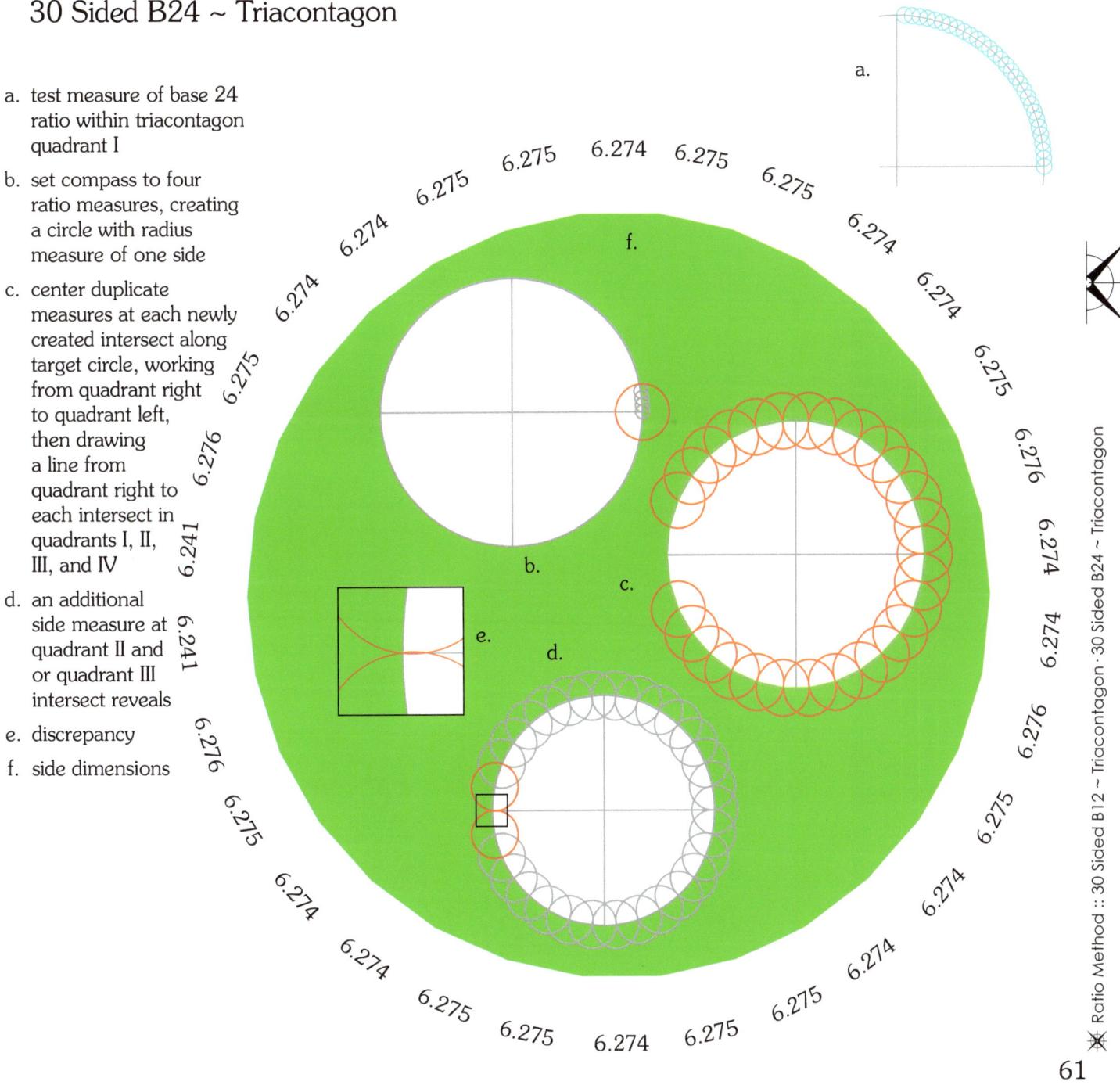

eXactagon Ratio Method

Color Stepped Composites

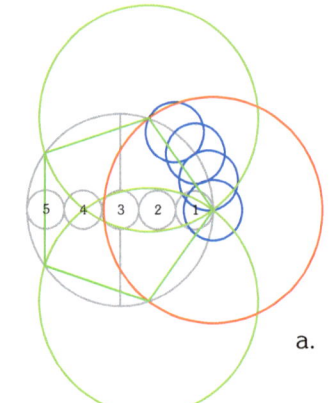

a.

b.

a. pentagon
b. heptagon
c. enneagon
d. hendecagon
e. tridecagon
f. pentadecagon
g. heptadecagon
h. enneadecagon

Step
Ordering

Later

Earlier

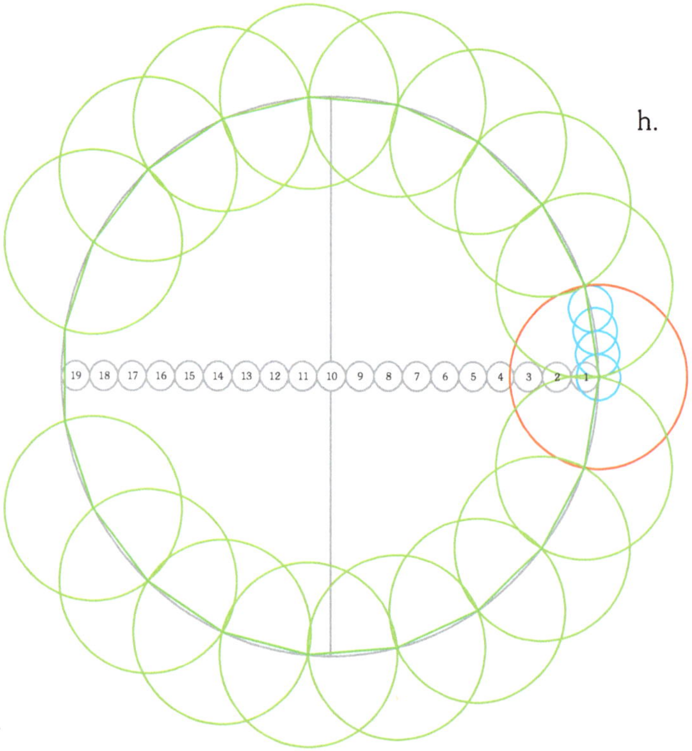

g.

h.

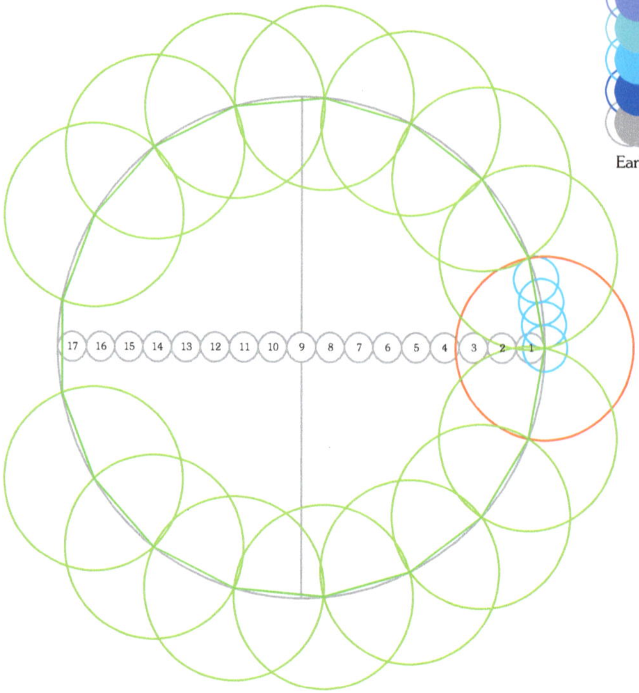

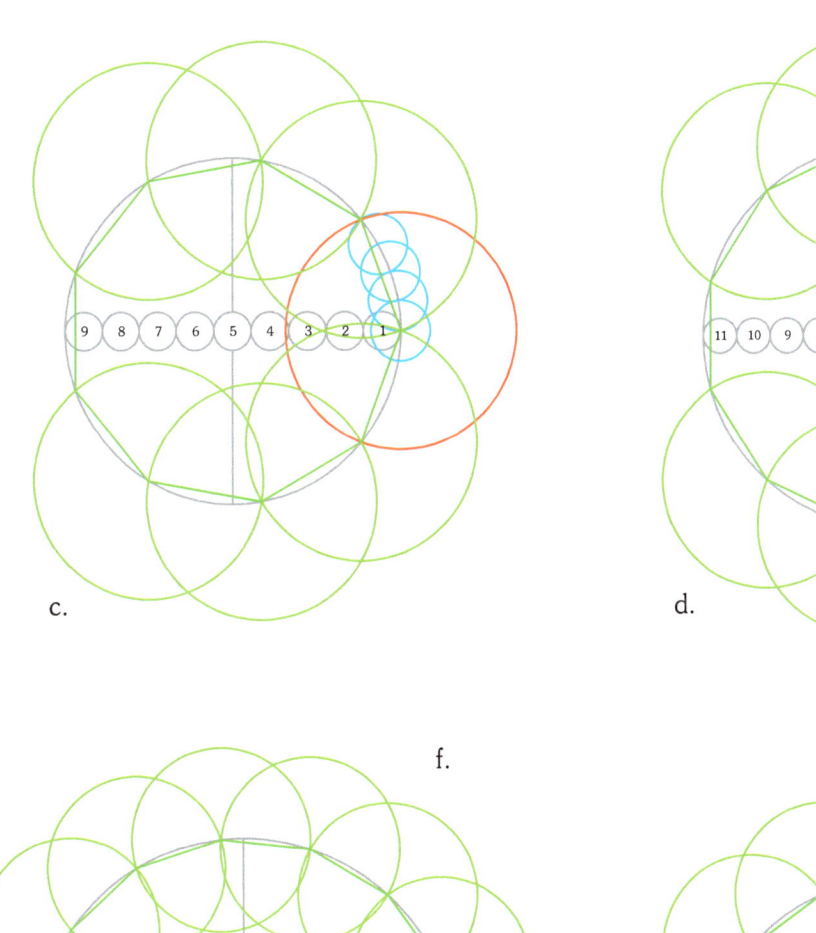

c.

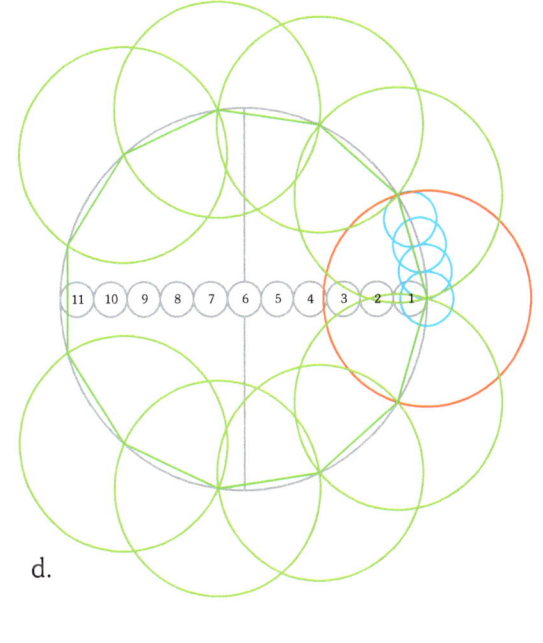

d.

f.

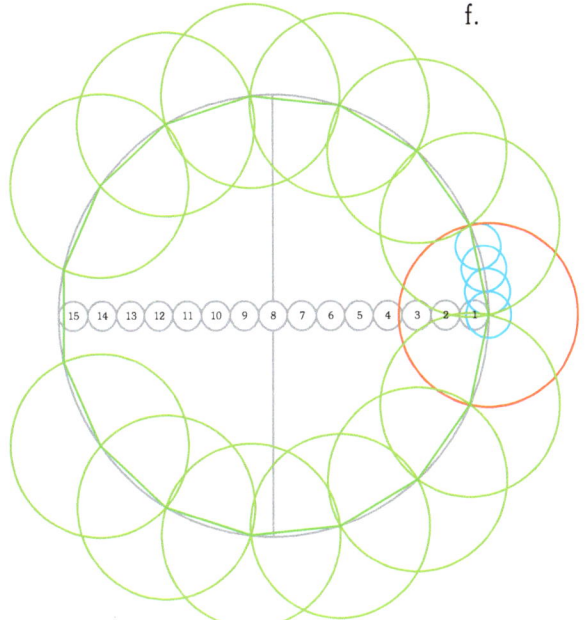

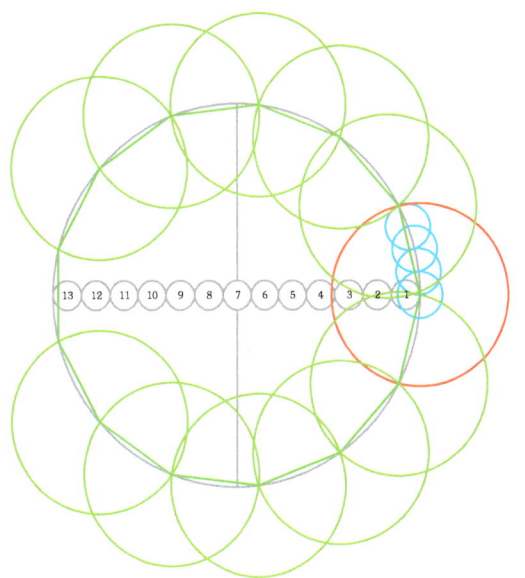

e.

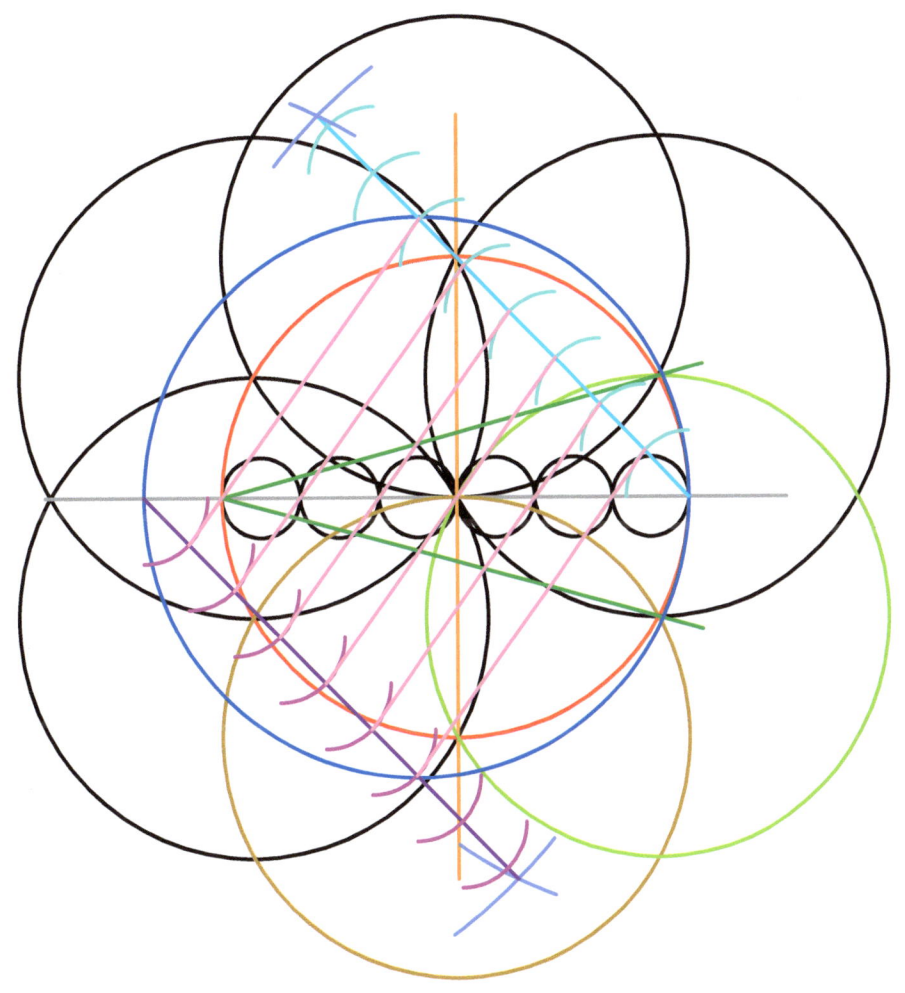

Scaling To Fit

Scaling To Fit

Scaling Simple

A simplified process of scaling measures to specific target size circles is to draw desired diameters within a given method of measure then obtain new related measures at appropriate intersections, placing newly realized measures onto a clean target circle.

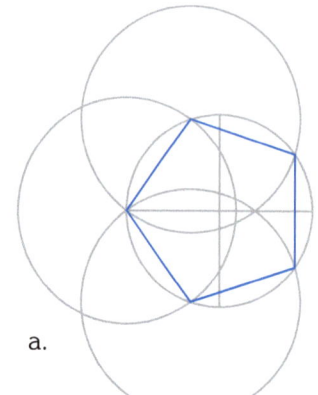

a.

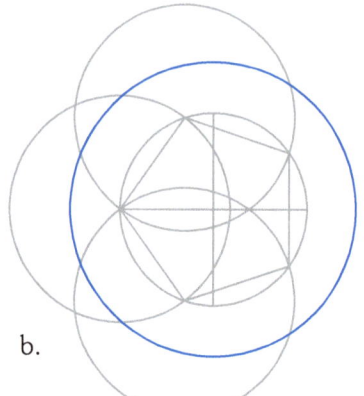

b.

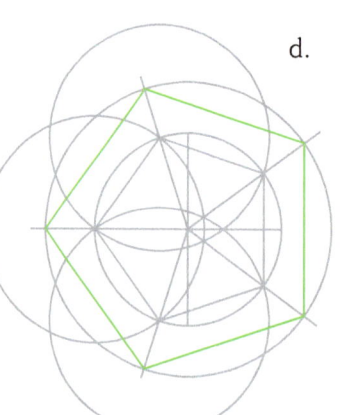

d.

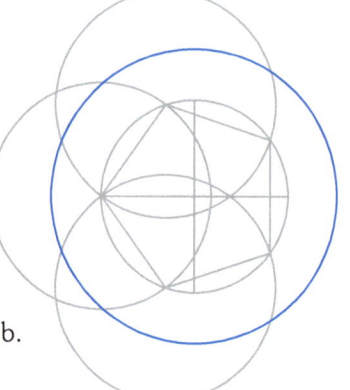

c.

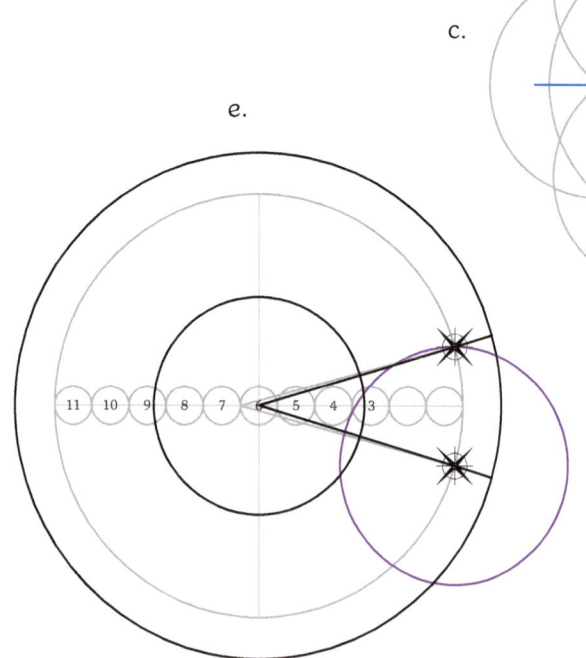

e.

Step Ordering

Later

Earlier

a. source polygon creation

b. place target circle at the identical origin of the source polygon circle

c. draw lines from origin through necessary points to intersect with target circle

d. draw target polygon

e. representation of two new target circles to be associated with the aligned base 11 circle

7 Sided Specific Setup

a. create target circle

b. draw an angled line from quadrant right, then divide equally

c. duplicate angled line from quadrant left; "Divide Line Equally" on page 14

d. divide new line identically from quadrant left

e. divide diameter into equal parts

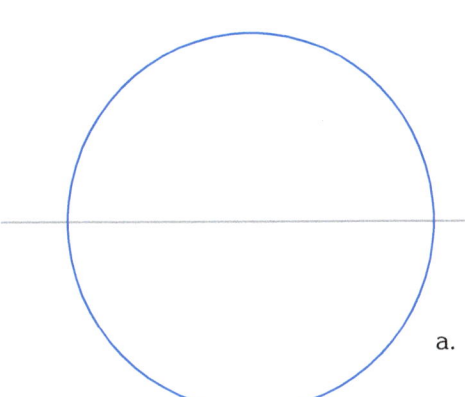

a.

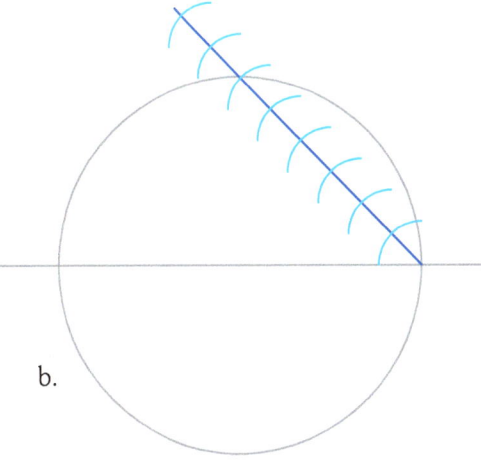

b.

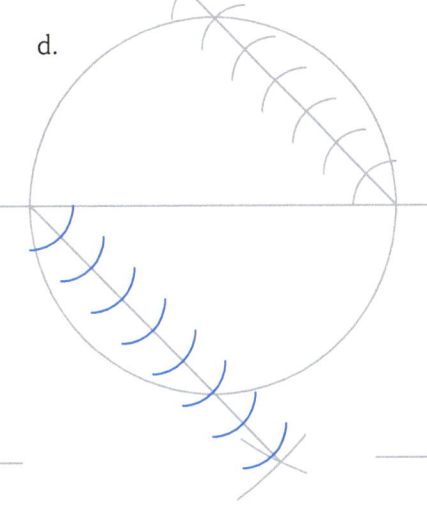

d.

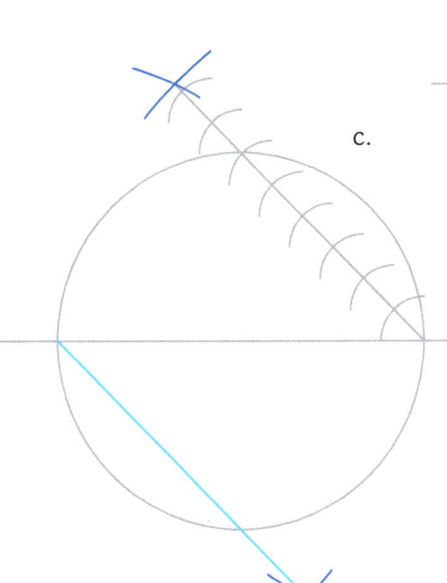

c.

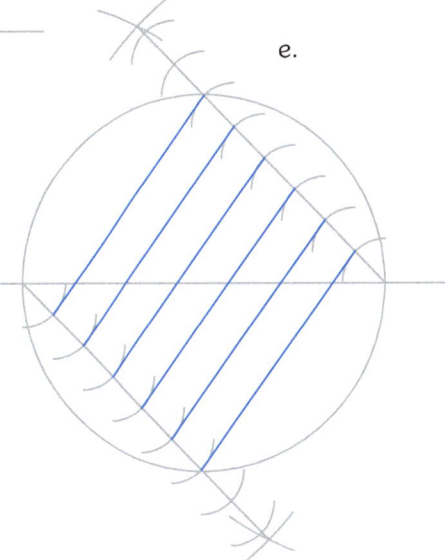

e.

Scaling To Fit

7 Sided ~ Align Base 6

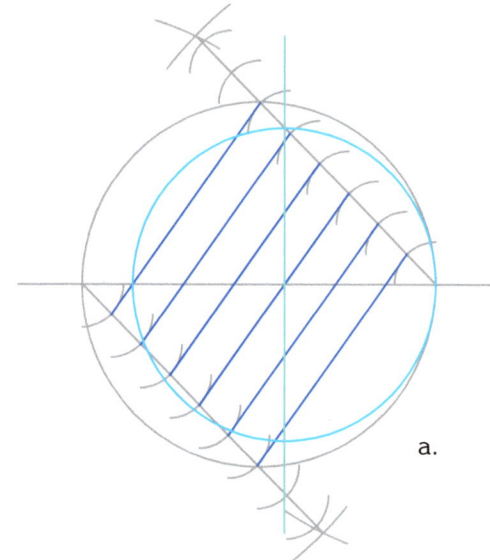

a.

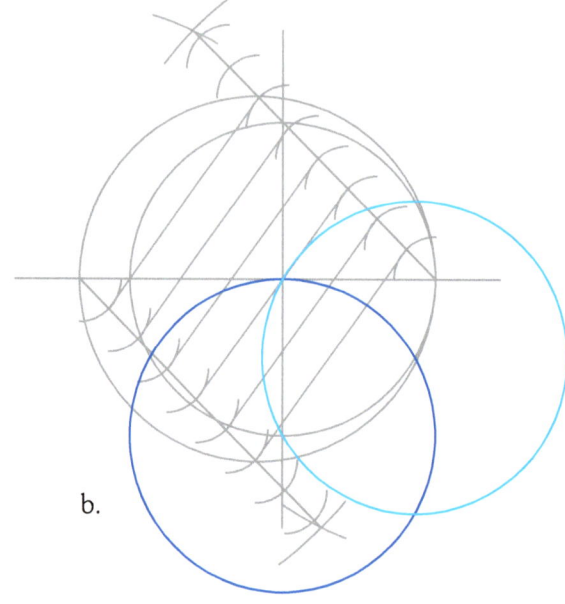

b.

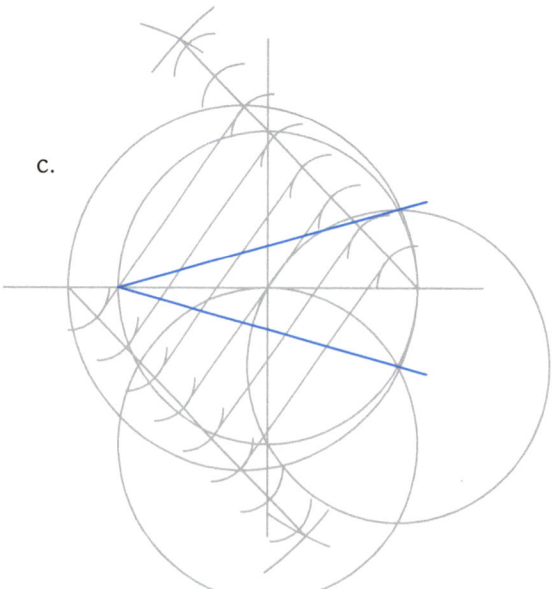

c.

Step
Ordering

Later

Earlier

a. continuing from
"7 Sided Specific
Setup" on page
67 create a
base 6 circle

b. duplicate circle at
quadrant bottom
then at quadrant IV
intersect

c. utilize "7 Sided B6
Arc ~ Heptagon" on
page 27

5 Sided ~ Align Base 6

a. divide target
   circle into equal
   segments

b. utilizing one
   segment measure

c. create base 6
   circle along with
   vertical axis,
   see"Circle With
   Center Axes" on
   page 13

d. duplicate circle of
   base 6 measure
   at quadrant
   bottom

e. continue
   duplication at
   quadrant IV
   intersect, then
   reference
   "5 Sided B6 Arc
   ~ Pentagon" on
   page 26

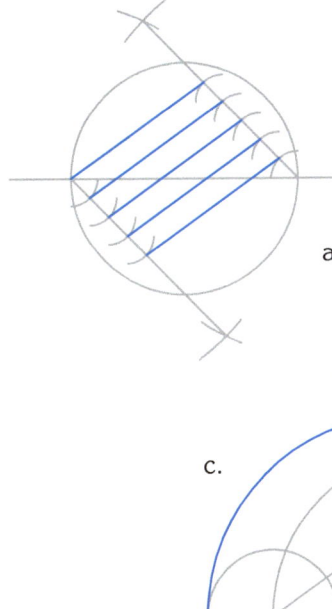

a.

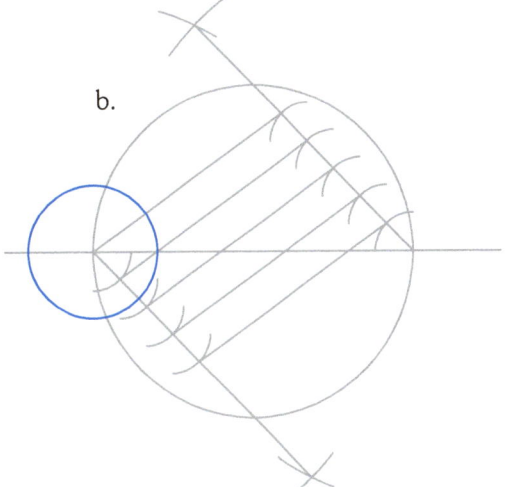

b.

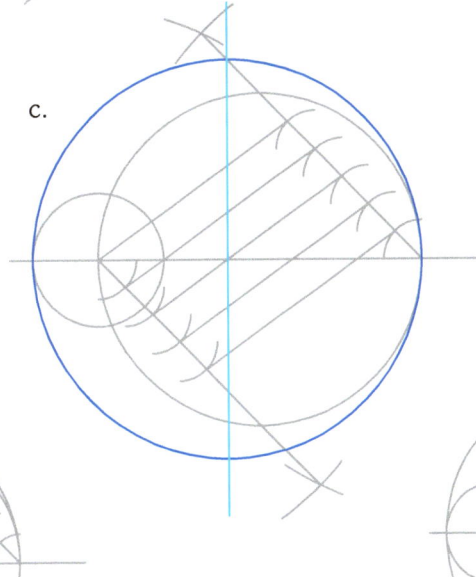

c.

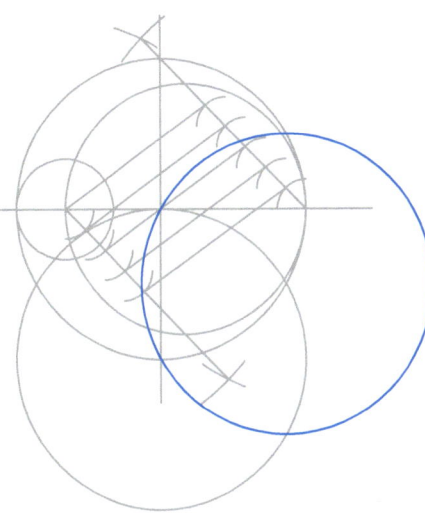

d.

e.

Scaling To Fit

7 Sided ~ Ratio Base 6

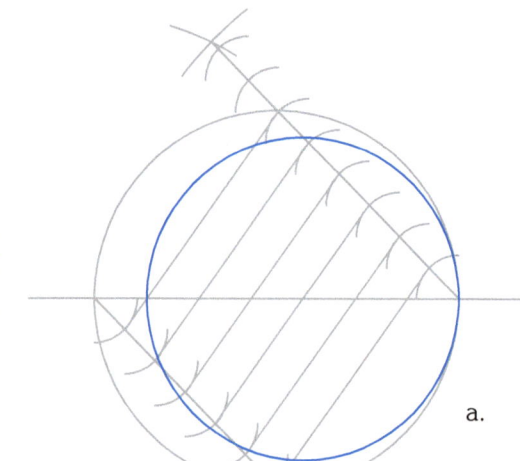

a.

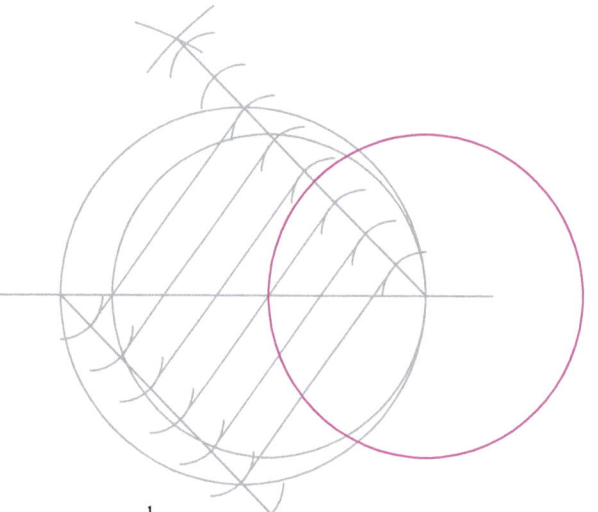

b.

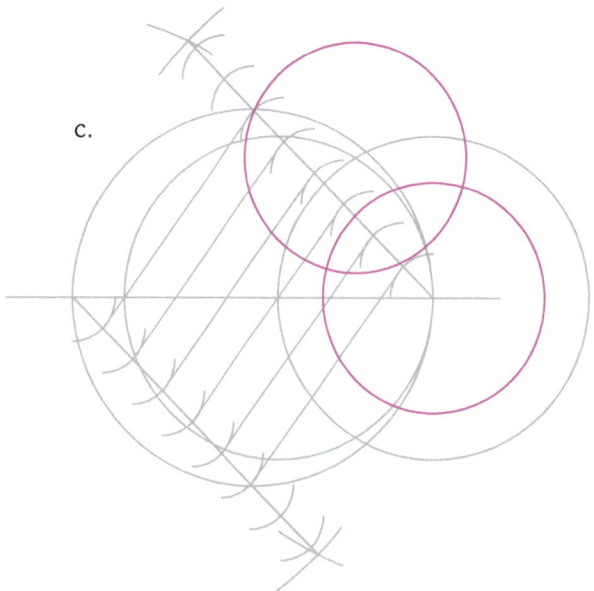

c.

Step
Ordering

Later

Earlier

a. continuing from "7 Sided Specific Setup" on page 67 create base 6 circle

b. duplicate base 6 circle measure at quadrant right to indicate arc in quadrant I to be used for base ratio

c. place identical circles at quadrant I intersect of base 6 and quadrant right

d. bisect quadrant I arc

e. place identical circles at bisect intersect and quadrant right

f. further bisect half arc

g. measure base 6 ratio, see "Base 6" on page 48 and "7 Sided B6 ~ Heptagon" on page 51

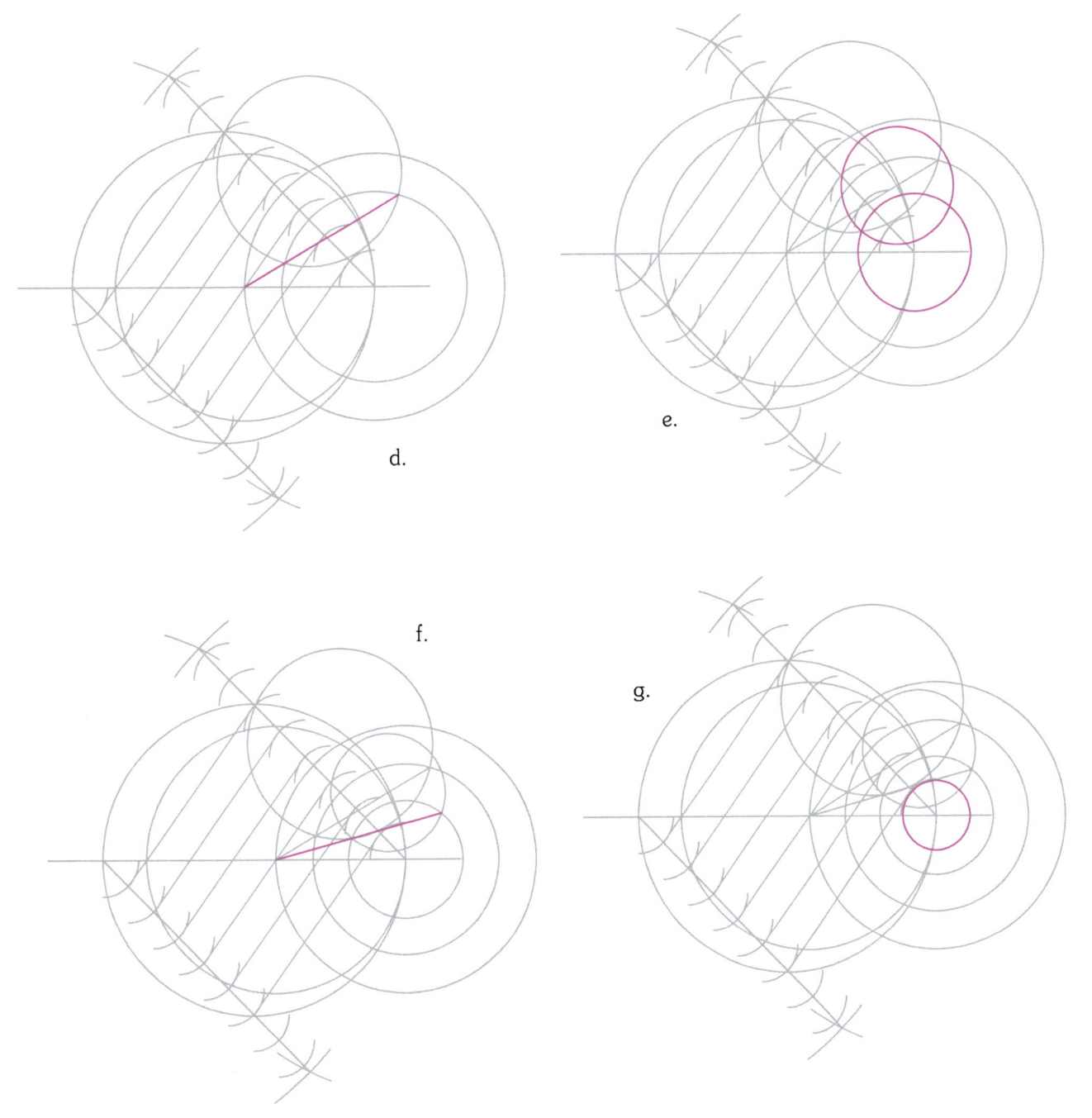

d.

e.

f.

g.

Color Stepped Composites

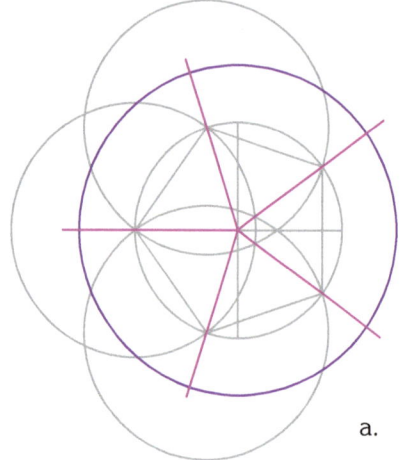

a.

a. simple pentagon

b. 5 sided align base 6

c. 7 sided align base 6

d. 7 sided ratio base 6

e. 11 sided align base 6

Step
Ordering

Later

Earlier

b.

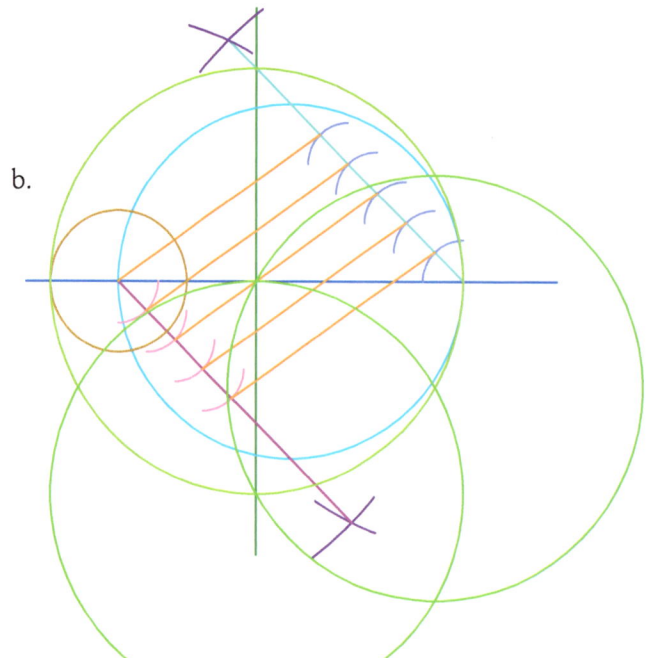

c.

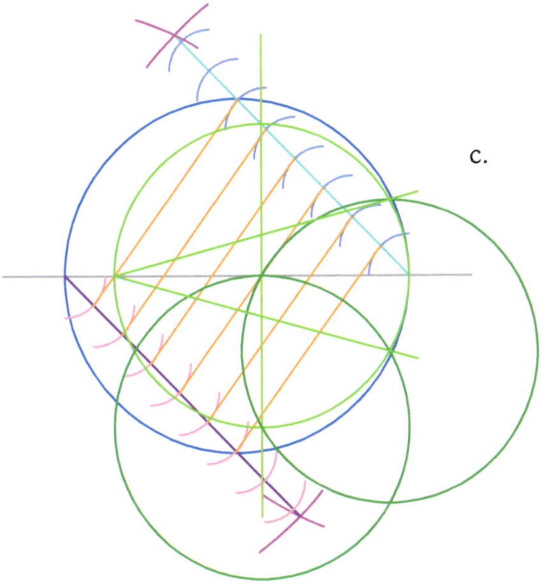

d.

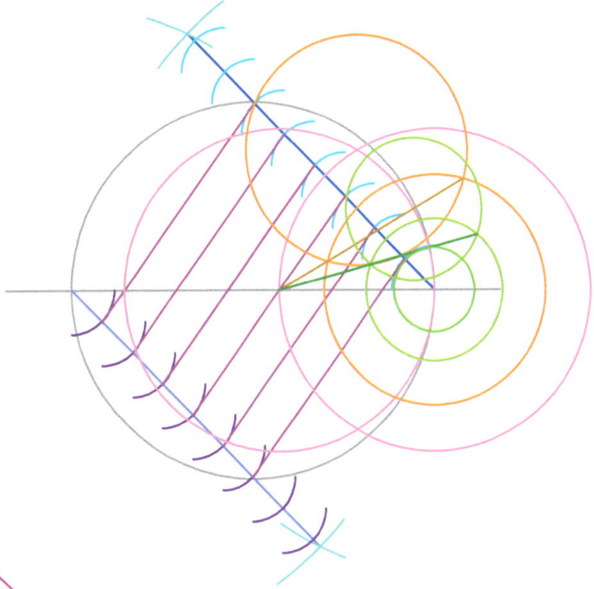

e.

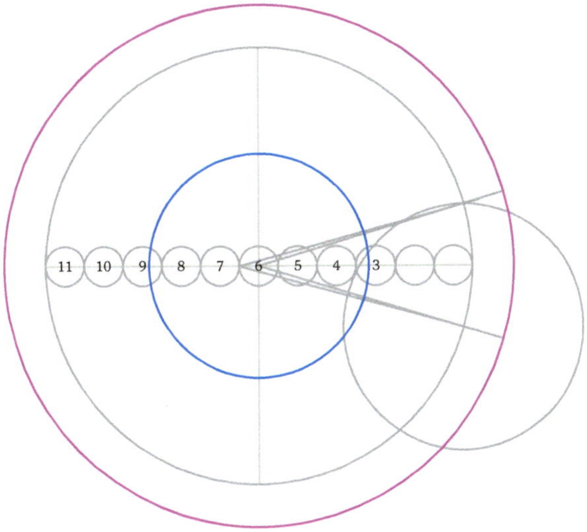

11 10 9 8 7 6 5 4 3

PERFECTAGØN

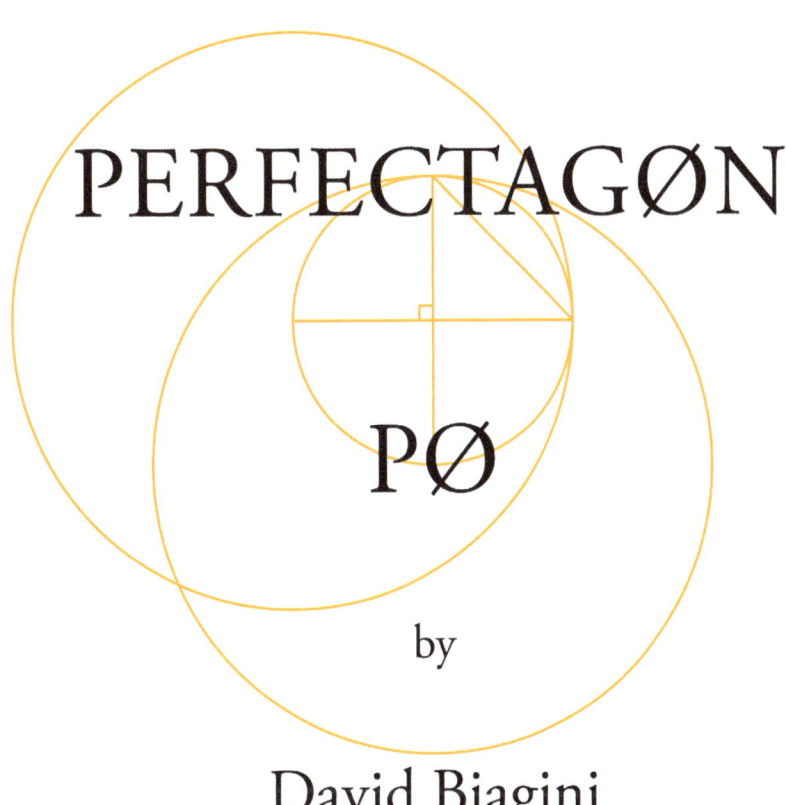

PØ

by

David Biagini

Perfectagon

Primer

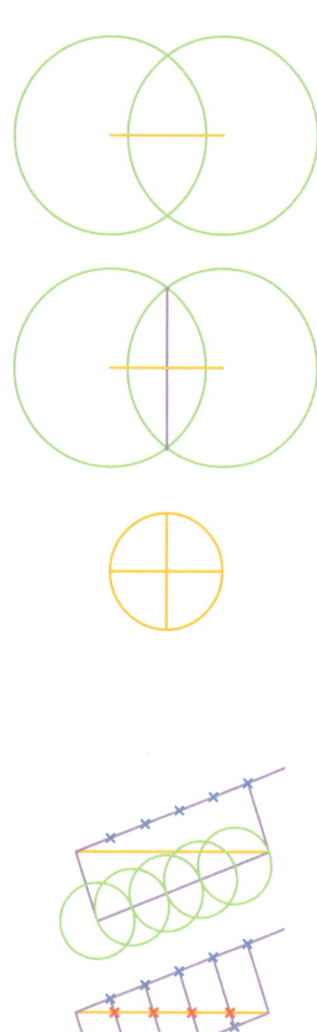

The Perfectagon Base Setup includes some standard practices for using a compass and straightedge to draw regular polygons. The additional steps that are standards for Perfectagon are the inclusion of a diagonal line across the upper right quadrant and two more circles. The additional circles are made by doubling the original circle's diameter, then placing the focal points at the lower endpoint and left endpoint of the cross. The placement of the additional circles creates an intersection that is used to help determine the side length for the target polygon.

An important aspect of Perfectagon is to use the compass and straightedge to equally divide a line into the number of segments equaling the number of sides that you wish your regular polygon to be. For instance divide a line into five equal segments for creating a pentagon. Dividing a line into equal segments can be a rather tedious step but it is an essential aspect of Perfectagon. An alternative to creating a segmented line is to use graph paper as a way to test the effectiveness of Perfectagon.

However the specific goal is to be able to create any regular polygon with only a compass and straightedge, therefore being able to divide a line into equal segments is an imperative step using Perfectagon. The illustrations to the left present some compass and straightedge standards that are required for using Perfectagon effectively. To the right is the Base Setup. The following pages reveal the steps for utilizing Perfectagon to draw regular polygons with 3, 4, 5, 6, 7, 8, 9, and 13 sides and ultimately any regular polygon.

Base Setup

Perfectagøn Base Setup

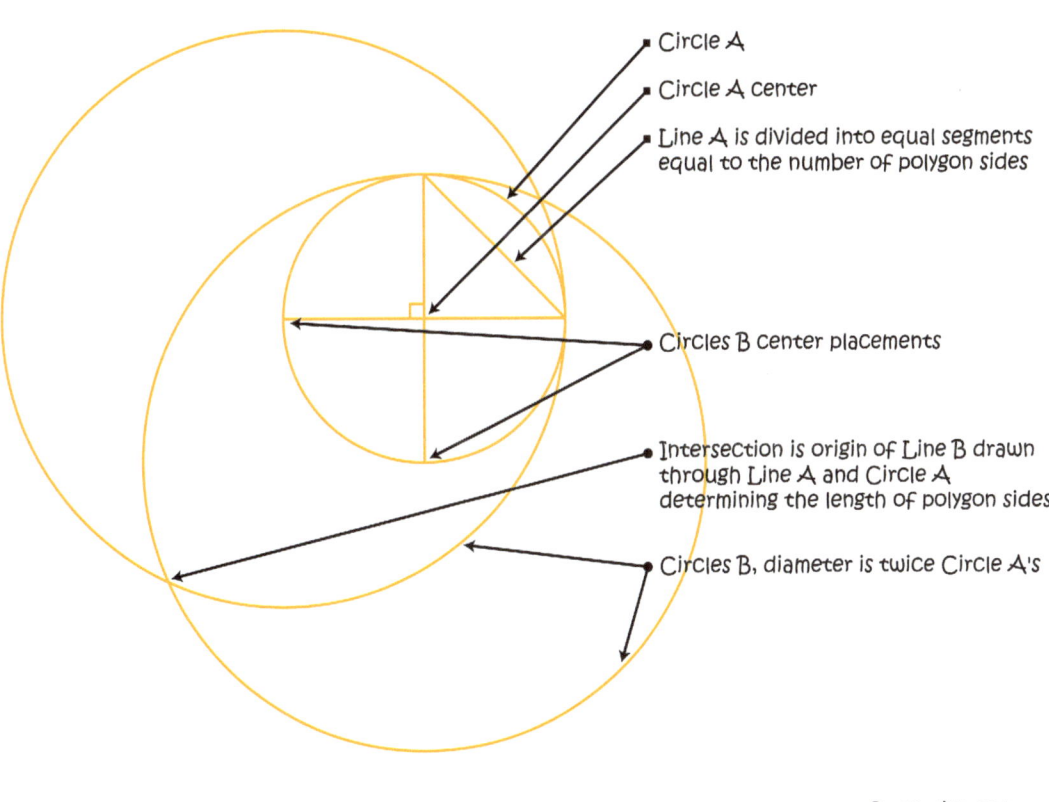

Circle A

Circle A center

Line A is divided into equal segments
equal to the number of polygon sides

Circles B center placements

Intersection is origin of Line B drawn
through Line A and Circle A
determining the length of polygon sides

Circles B, diameter is twice Circle A's

Imovative.com

3 Sided ~ Trigon

Perfectagøn 3 Sided

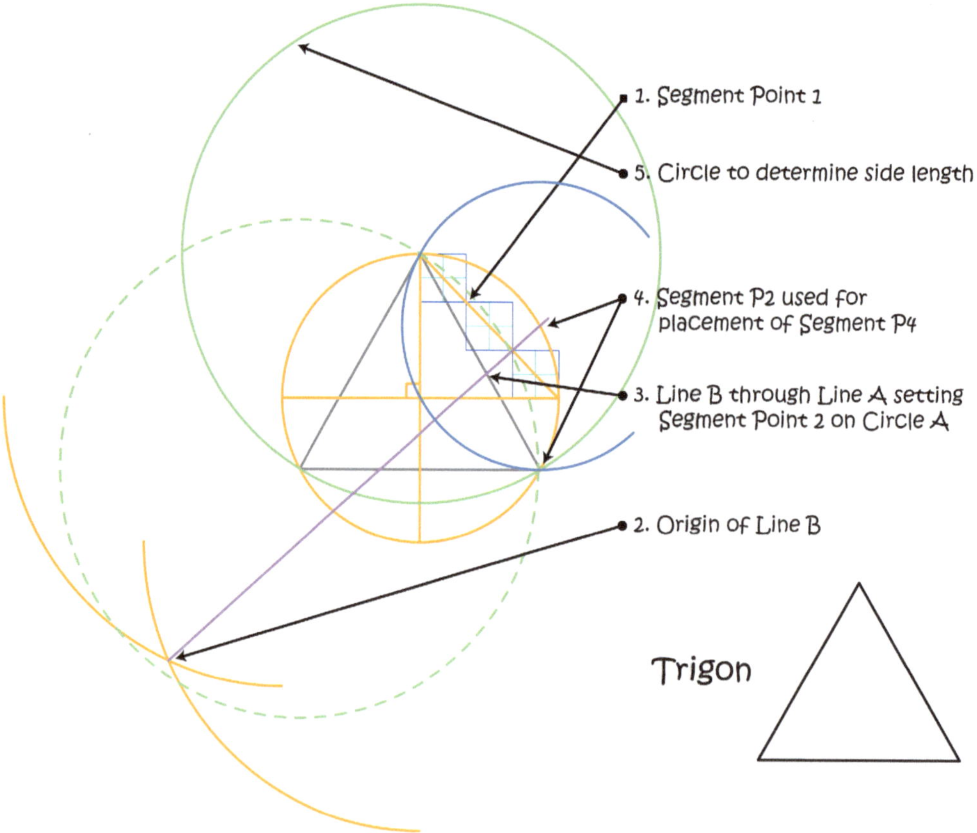

1. Segment Point 1

5. Circle to determine side length

4. Segment P2 used for placement of Segment P4

3. Line B through Line A setting Segment Point 2 on Circle A

2. Origin of Line B

Trigon

4 Sided ~ Tetragon

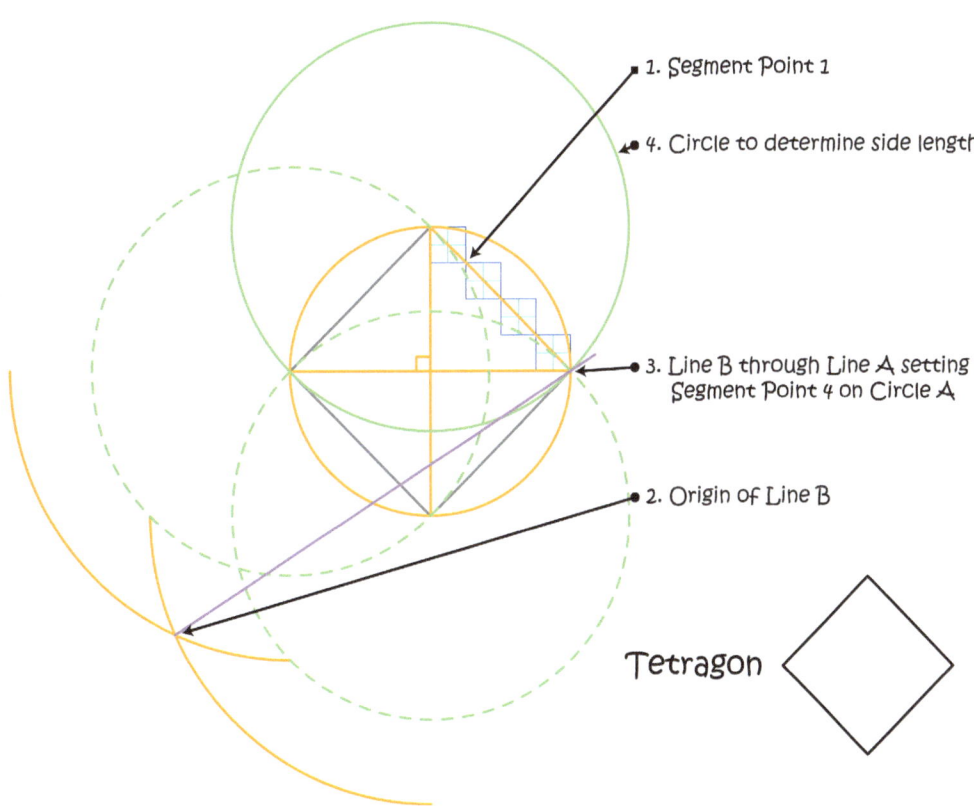

Perfectagøn 4 Sided

1. Segment Point 1

4. Circle to determine side length

3. Line B through Line A setting Segment Point 4 on Circle A

2. Origin of Line B

Tetragon

5 Sided ~ Pentagon

Perfectagøn 5 Sided

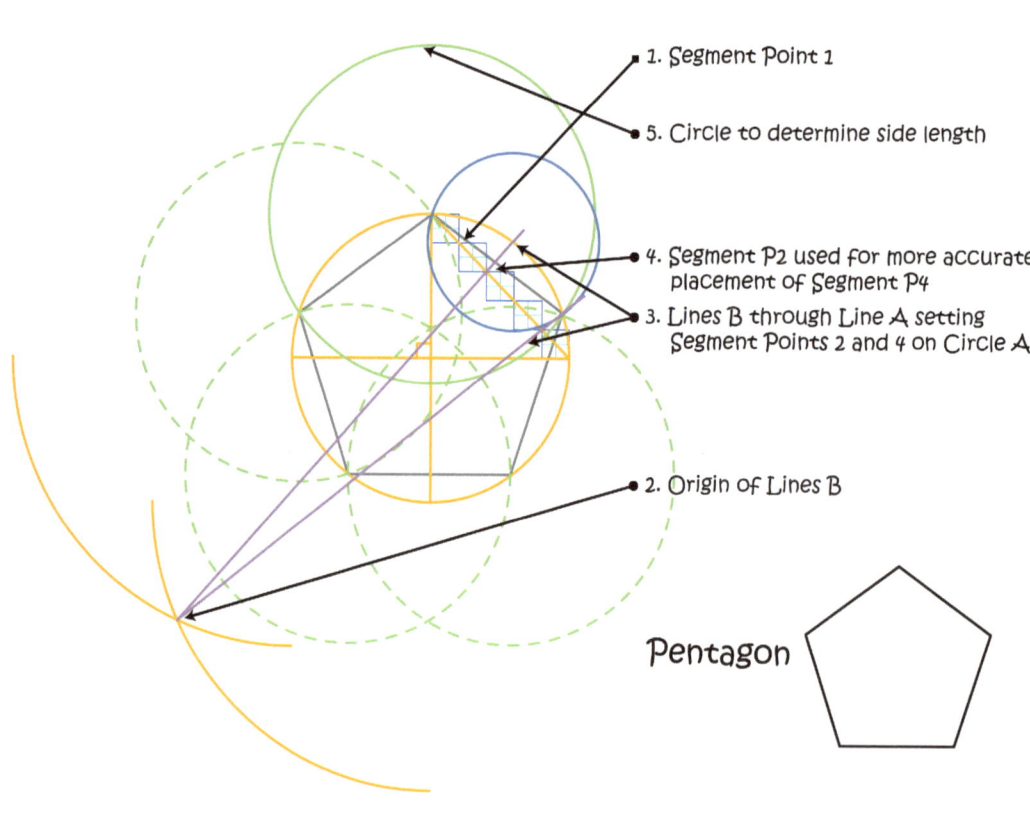

1. Segment Point 1

5. Circle to determine side length

4. Segment P2 used for more accurate placement of Segment P4

3. Lines B through Line A setting Segment Points 2 and 4 on Circle A

2. Origin of Lines B

Pentagon

6 Sided ~ Hexagon

Perfectagøn 6 Sided

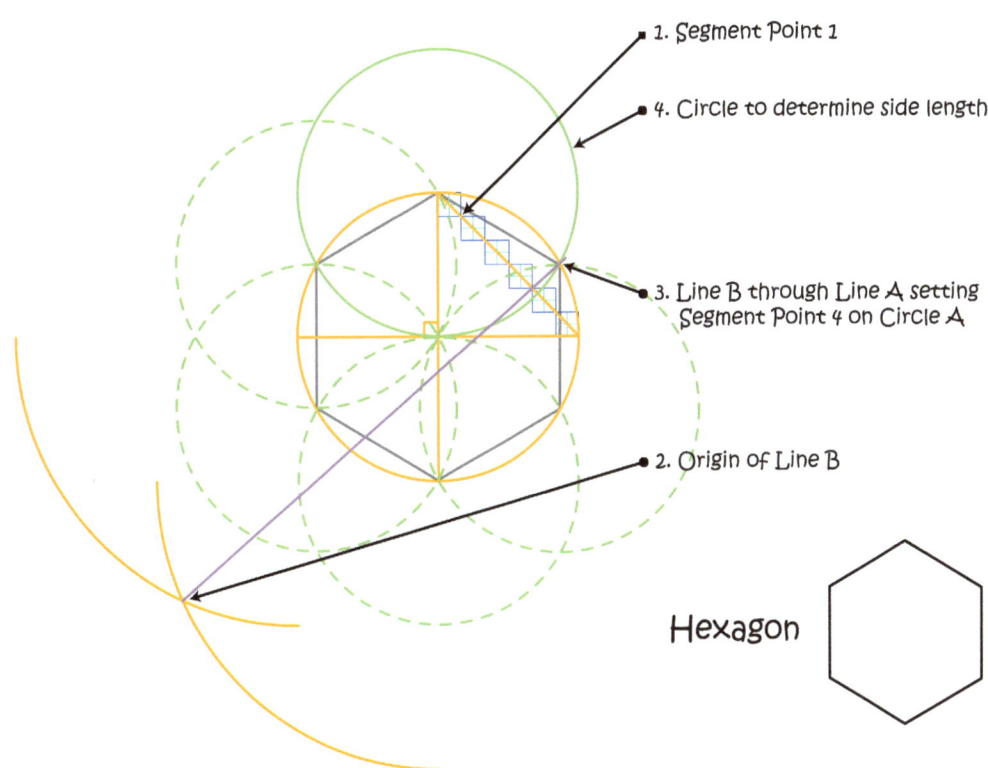

1. Segment Point 1

4. Circle to determine side length

3. Line B through Line A setting
Segment Point 4 on Circle A

2. Origin of Line B

Hexagon

Perfectagon

7 Sided ~ Heptagon

Perfectagøn 7 Sided

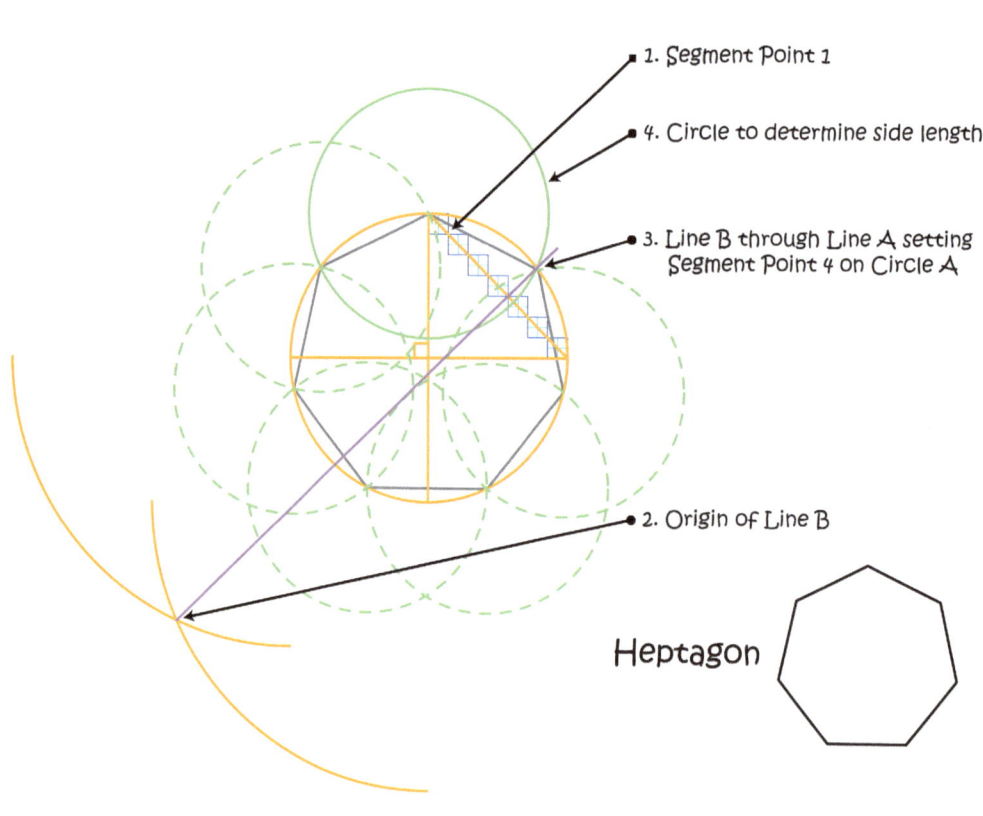

1. Segment Point 1

4. Circle to determine side length

3. Line B through Line A setting Segment Point 4 on Circle A

2. Origin of Line B

Heptagon

8 Sided ~ Octagon

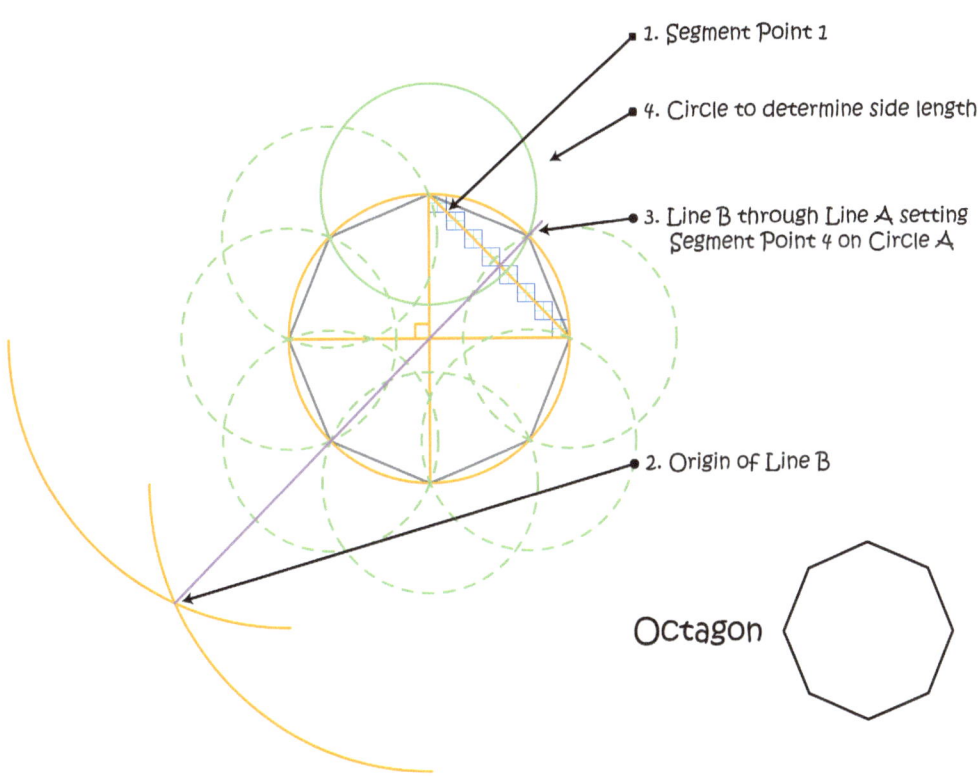

Perfectagøn 8 Sided

1. Segment Point 1

4. Circle to determine side length

3. Line B through Line A setting Segment Point 4 on Circle A

2. Origin of Line B

Octagon

9 Sided ~ Enneagon

Perfectagøn 9 Sided

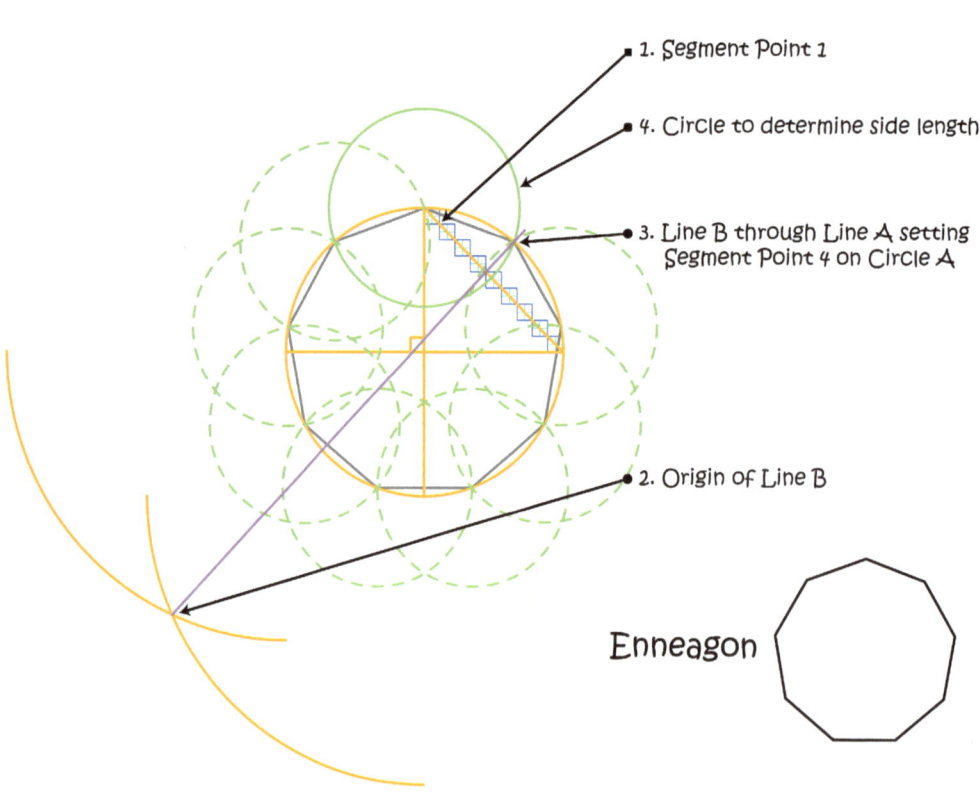

1. Segment Point 1

4. Circle to determine side length

3. Line B through Line A setting
   Segment Point 4 on Circle A

2. Origin of Line B

Enneagon

13 Sided ~ Tridecagon

Perfectagøn 13 Sided

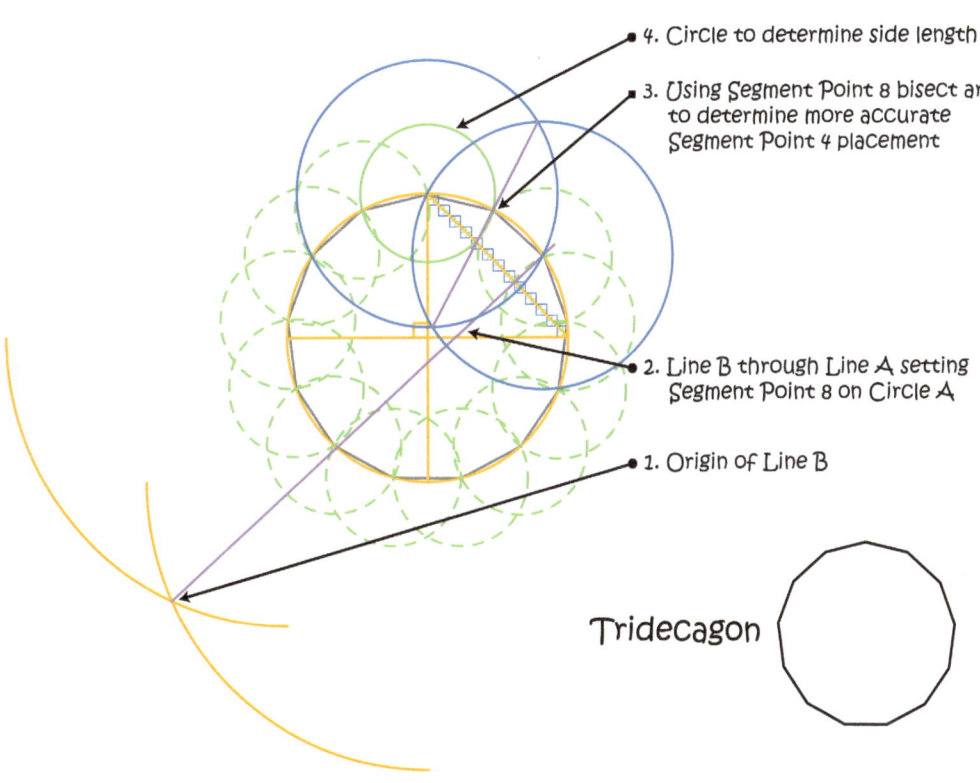

4. Circle to determine side length

3. Using Segment Point 8 bisect arc to determine more accurate Segment Point 4 placement

2. Line B through Line A setting Segment Point 8 on Circle A

1. Origin of Line B

Tridecagon

Puzzle Postage

Perfectagon Stamps

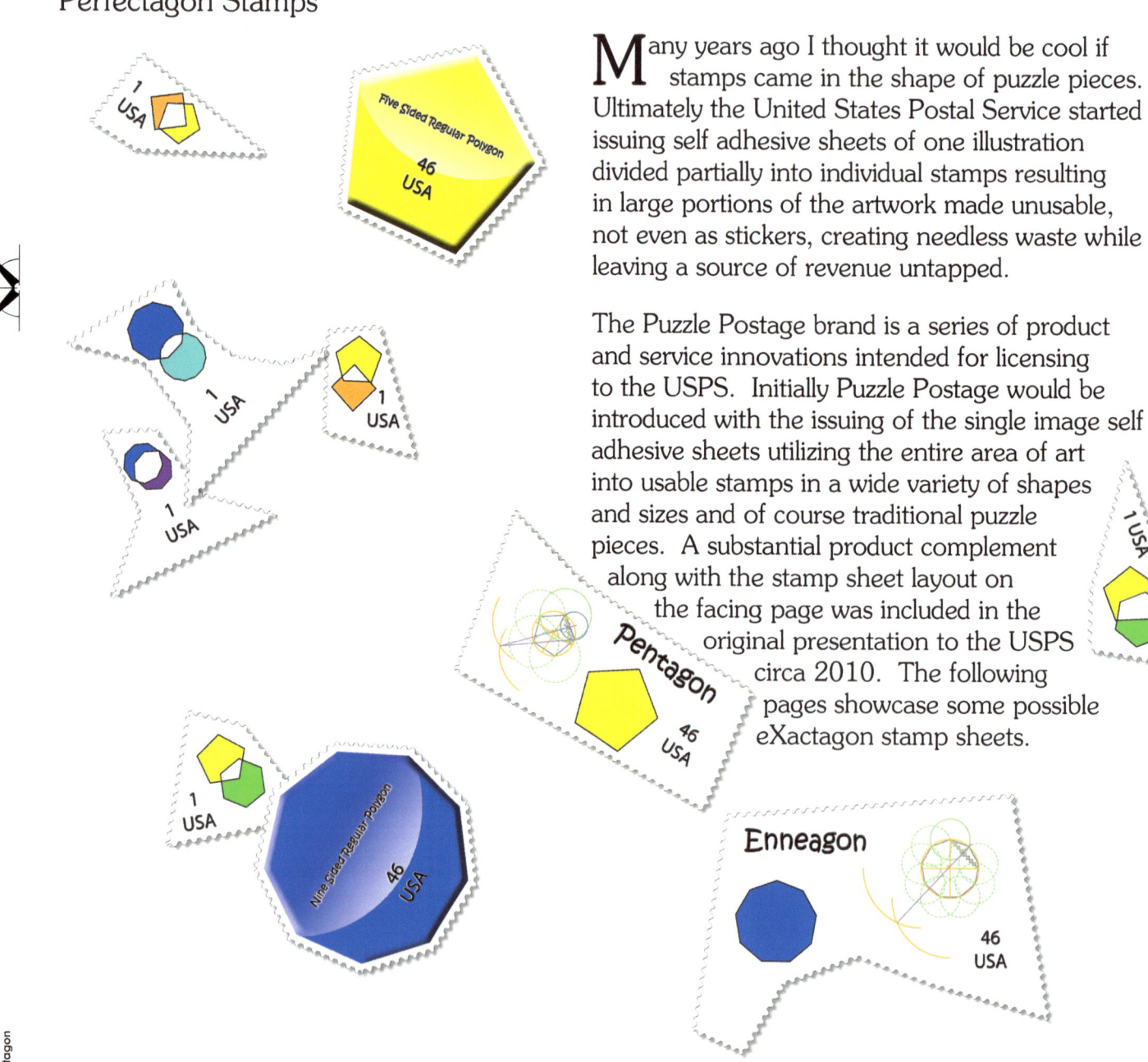

M any years ago I thought it would be cool if stamps came in the shape of puzzle pieces. Ultimately the United States Postal Service started issuing self adhesive sheets of one illustration divided partially into individual stamps resulting in large portions of the artwork made unusable, not even as stickers, creating needless waste while leaving a source of revenue untapped.

The Puzzle Postage brand is a series of product and service innovations intended for licensing to the USPS. Initially Puzzle Postage would be introduced with the issuing of the single image self adhesive sheets utilizing the entire area of art into usable stamps in a wide variety of shapes and sizes and of course traditional puzzle pieces. A substantial product complement along with the stamp sheet layout on the facing page was included in the original presentation to the USPS circa 2010. The following pages showcase some possible eXactagon stamp sheets.

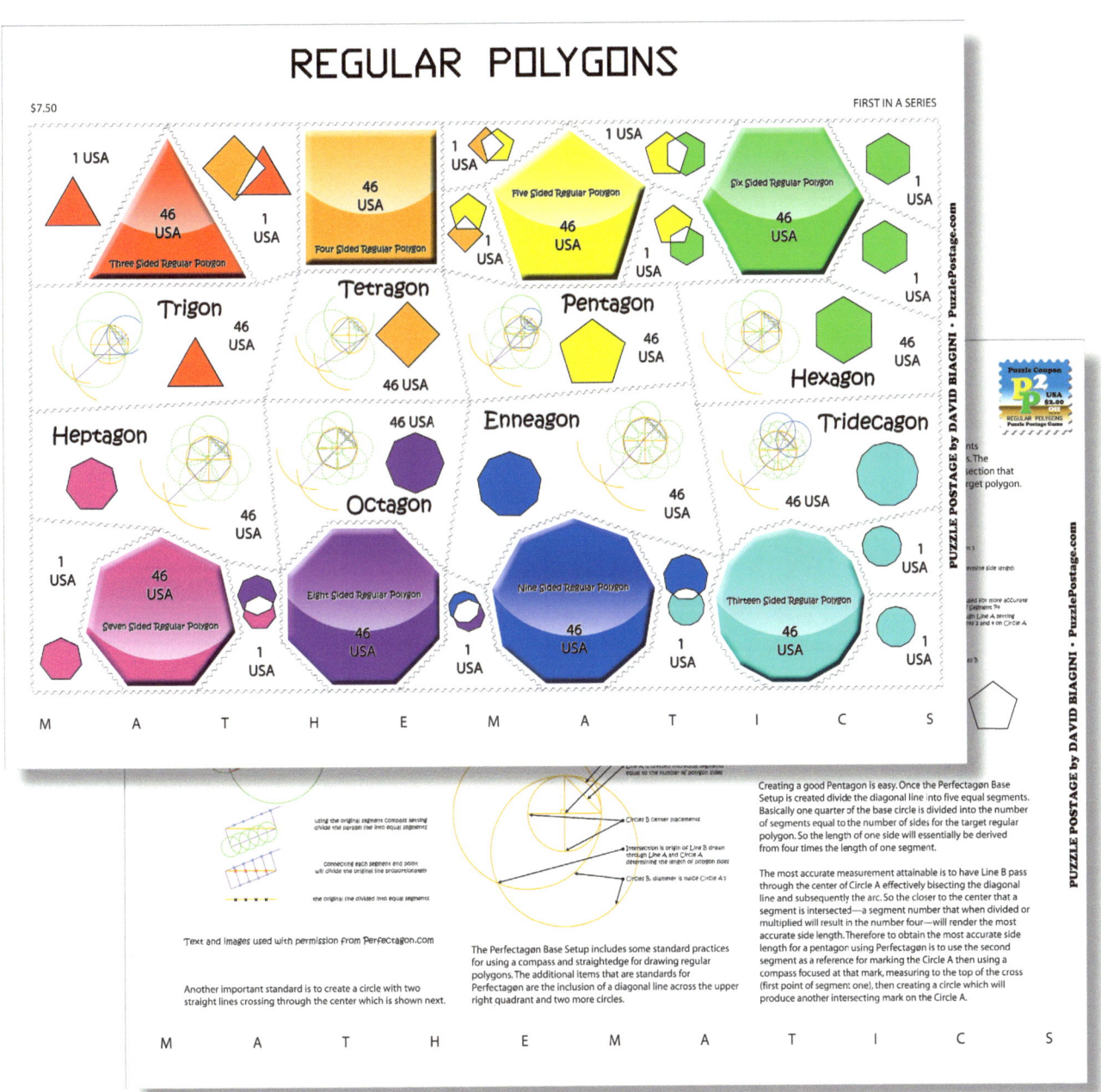

REGULAR POLYGONS

$7.50

Trigon — Three Sided Regular Polygon

Tetragon — Four Sided Regular Polygon

Pentagon — Five Sided Regular Polygon

Hexagon — Six Sided Regular Polygon

Heptagon — Seven Sided Regular Polygon

Octagon — Eight Sided Regular Polygon

Enneagon — Nine Sided Regular Polygon

Tridecagon — Thirteen Sided Regular Polygon

46 USA · 1 USA

M A T H E M A T I C S

PUZZLE POSTAGE by DAVID BIAGINI · PuzzlePostage.com

---

using the original segments compass seeing divide the parallel line into equal segments

connecting each segment and point will divide the original line proportionately

the original line divided into equal segments

Text and images used with permission from Perfectagon.com

Another important standard is to create a circle with two straight lines crossing through the center which is shown next.

The Perfectagon Base Setup includes some standard practices for using a compass and straightedge for drawing regular polygons. The additional items that are standards for Perfectagon are the inclusion of a diagonal line across the upper right quadrant and two more circles.

Circle B center placement

Intersection is origin of Line B drawn through Line A and Circle A, determining the length of polygon sides

Circle B diameter is inside Circle A

Creating a good Pentagon is easy. Once the Perfectagon Base Setup is created divide the diagonal line into five equal segments. Basically one quarter of the base circle is divided into the number of segments equal to the number of sides for the target regular polygon. So the length of one side will essentially be derived from four times the length of one segment.

The most accurate measurement attainable is to have Line B pass through the center of Circle A effectively bisecting the diagonal line and subsequently—a segment number that when divided or multiplied will result in the number four—will render the most accurate side length. Therefore to obtain the most accurate side length for a pentagon using Perfectagon is to use the second segment as a reference for marking the Circle A then using a compass focused at that mark, measuring to the top of the cross (first point of segment: one), then creating a circle which will produce another intersecting mark on the Circle A.

M A T H E M A T I C S

PUZZLE POSTAGE by DAVID BIAGINI · PuzzlePostage.com

eXactagon Align Forever Stamps

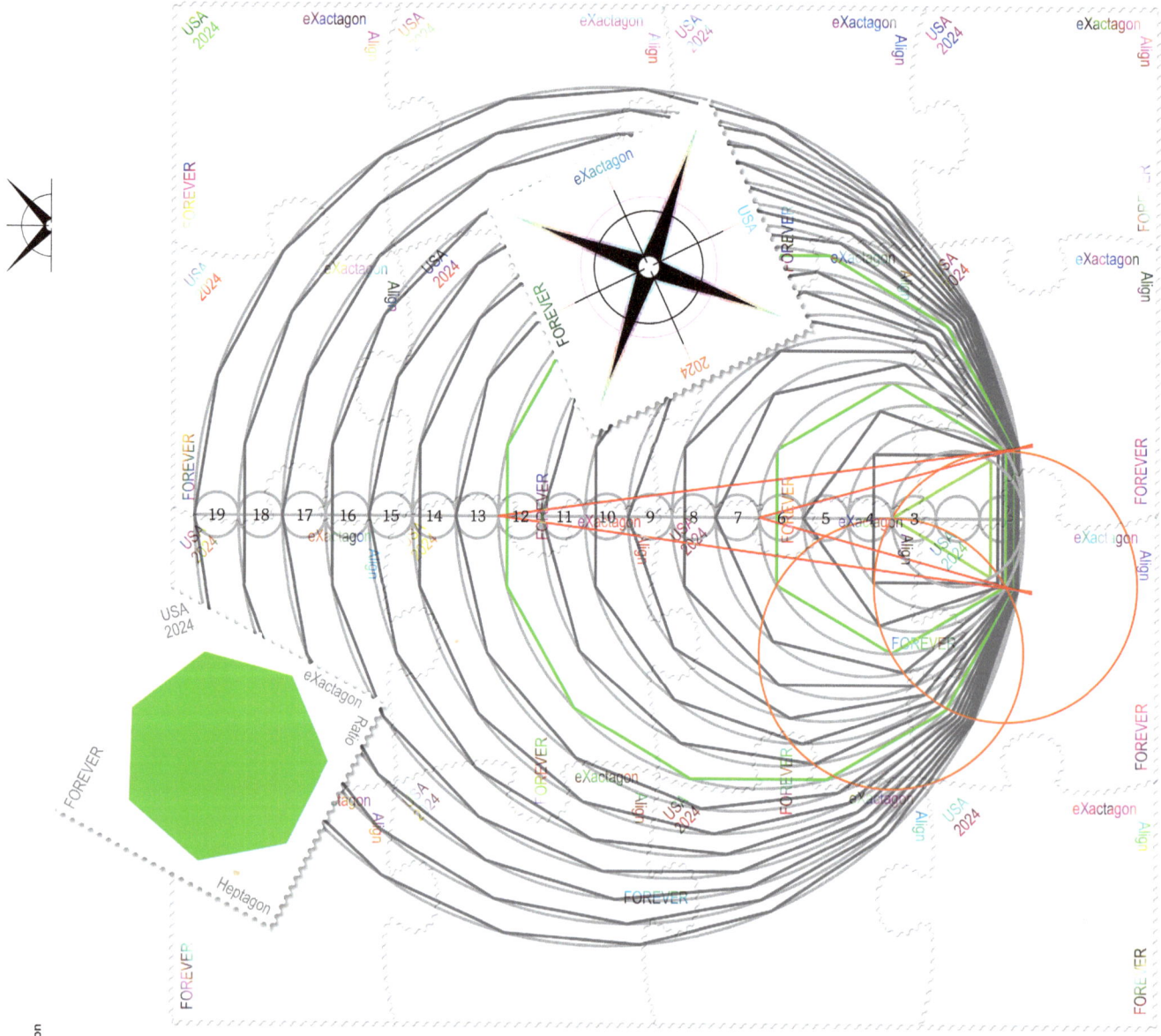

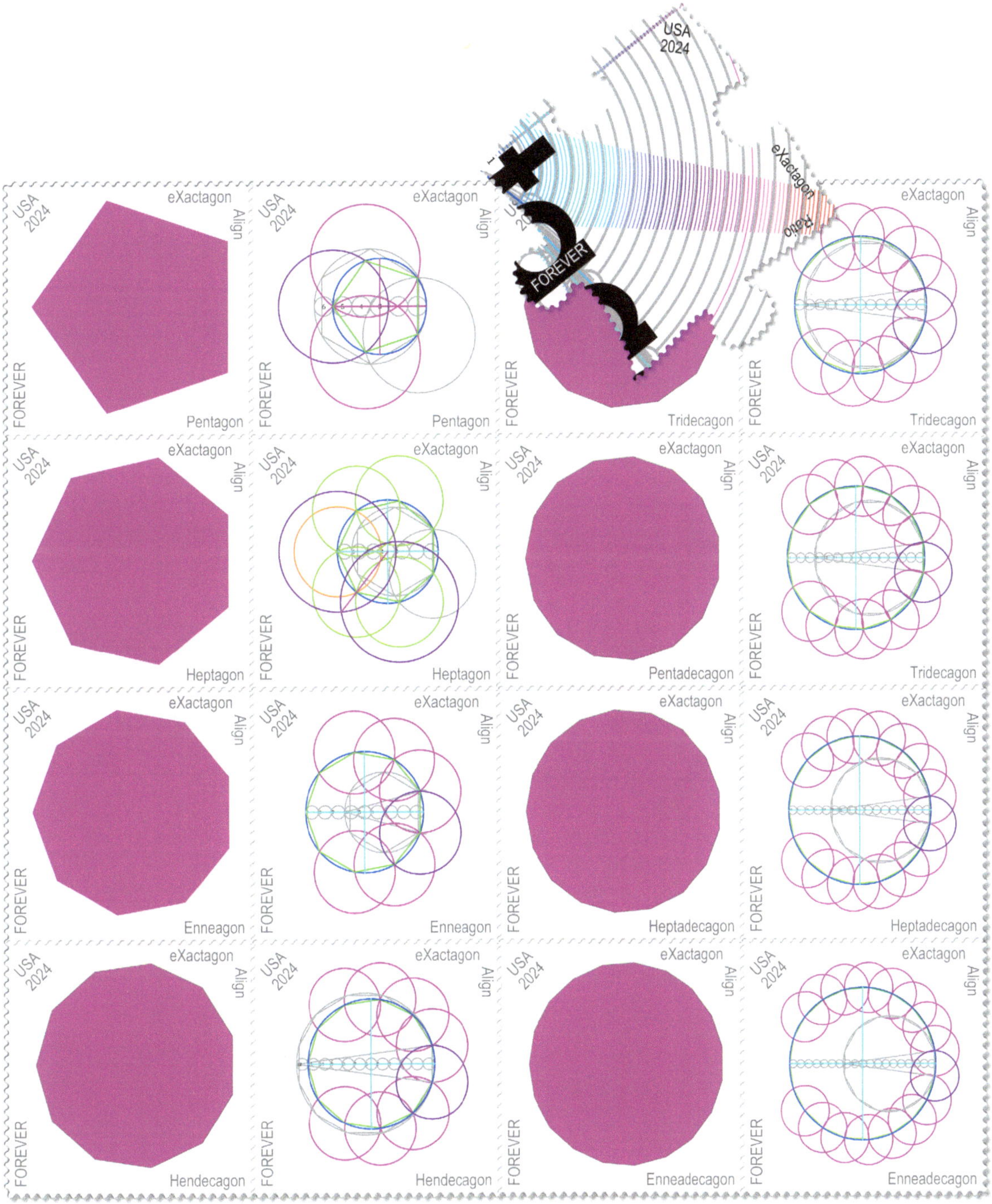

Puzzle Postage

eXactagon Ratio Forever Stamps

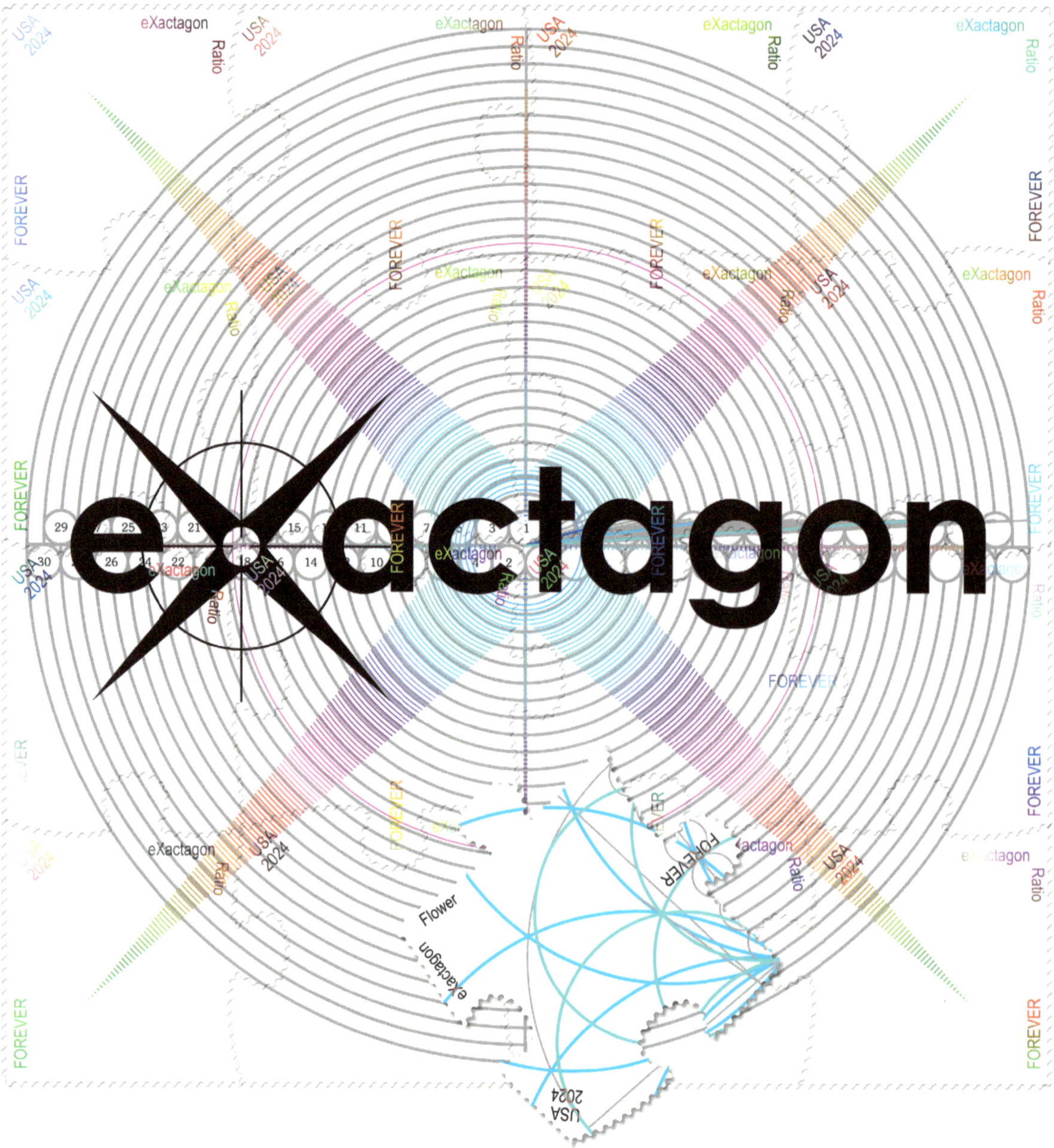

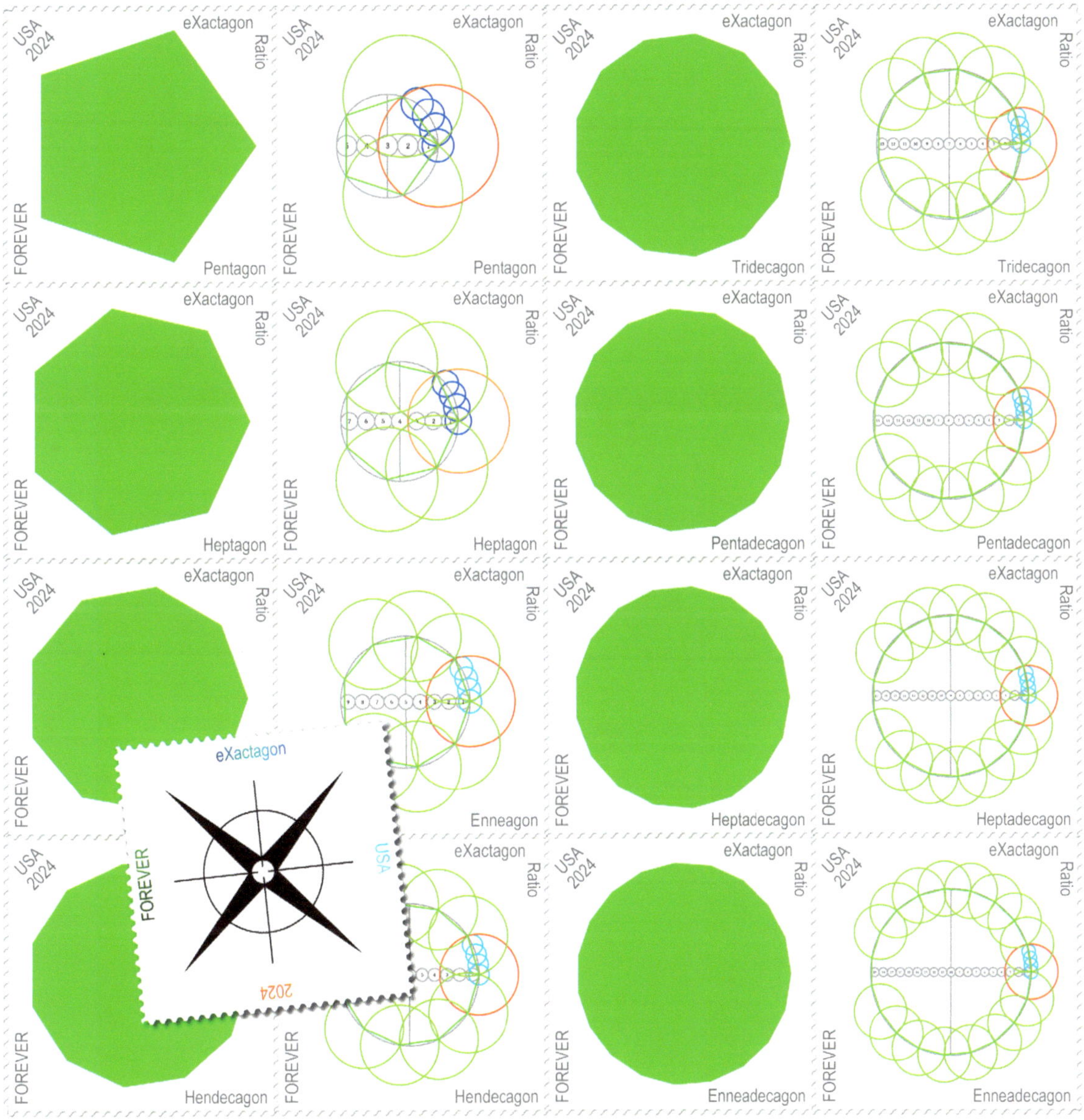

eXactagon Flower Forever Stamps

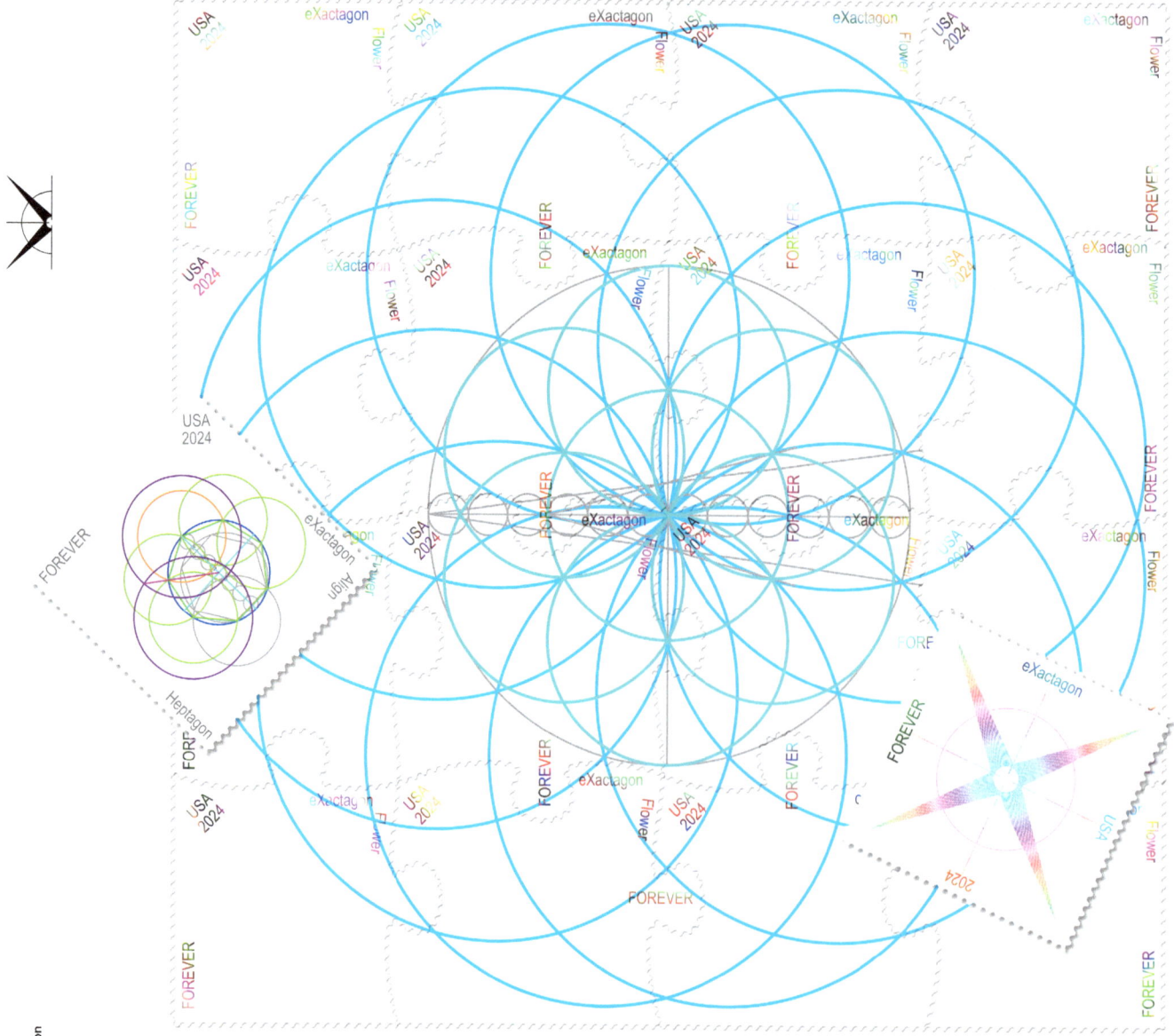

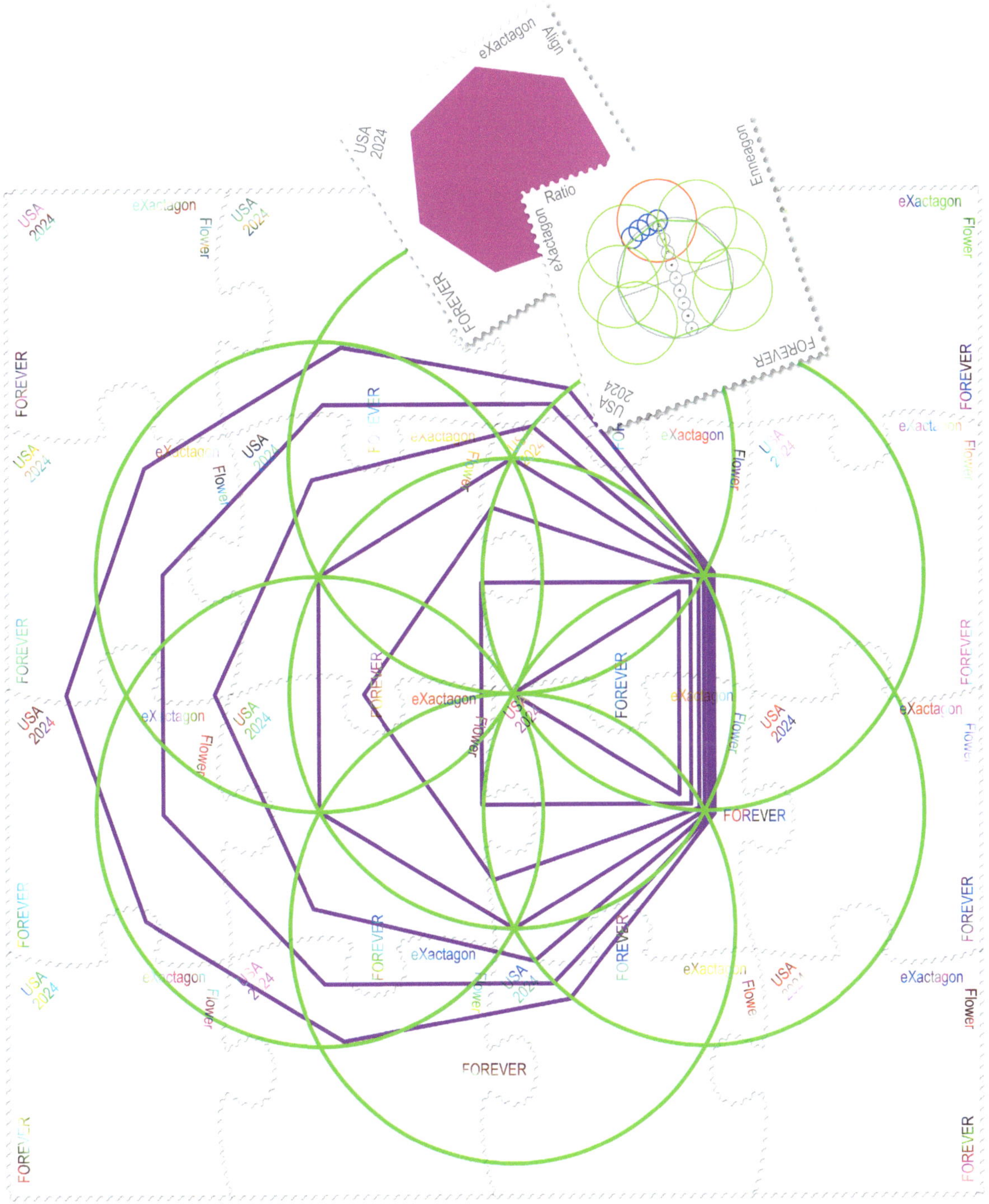
eXactagon
Align
USA 2024
FOREVER
Ratio
Enneagon
Flower

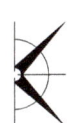

Other Titles

Trademarks

along with associated symbols
and certain phrases
throughout presentation

eXactagon
Perfectagon
Coloring Zen
Puzzle Postage
Thoughtstrip
Imovative
VOID Art
Clash Flow
Store Wears
Episode Fore
Money Charms
Sketch Look Book
Hello Hummingbirds
David Biagini
VOID